CAN YOU PLAY
CRICKET
ON MARS?

CAN YOU PLAY CRICKET ON MARS?

AND OTHER SCIENTIFIC
QUESTIONS ANSWERED

PATRICK MOORE

First published 2008

The History Press
The Mill, Brimscombe Port
Stroud, Gloucestershire, GL5 2QG
www.thehistorypress.co.uk

British Library Cataloguing in Publication Data.
A catalogue record for this book is available from the British Library.

ISBN 978 0 7509 5114 2

Typesetting and origination by The History Press
Printed in Great Britain

INTRODUCTION

This is a book of 'question and answer'. I will ask the questions, and do my best to give the answers. But at the very outset, I think it may be useful to give a brief rundown of the various components of the universe, introducing terms which will crop up time and time again.

The Earth on which we live is a planet, moving round the Sun in a period of one year. It is not the only one in the Sun's family; there are seven others, and these are the main members of the Solar System. A planet has no light of its own, and shines only by reflected sunlight. Most of the planets have secondary bodies or satellites moving round them; we have one satellite, the Moon, which also shines by reflected sunlight. The largest planets, Jupiter and Saturn, have over sixty satellites each, though most of them are very small.

Reckoning outward from the Sun, we come first to rocky, comparatively small planets: Mercury, Venus, the Earth and Mars. Then comes a wide gap, in which move thousands of small bodies known as asteroids. Beyond the asteroid belt come the four giant planets Jupiter, Saturn, Uranus and Neptune, with their satellite families; the giants are not rocky, but have gaseous surfaces. Beyond the path (orbit) of Neptune lies another swarm of smaller bodies, making up the Kuiper Belt; the largest members of which are Eris and Pluto. Pluto, the first member of the swarm to be discovered (in 1930) was long classed as a planet, but has now been officially relegated to the status of an ordinary Kuiper Belt Object or KBO.

Comets also move round the Sun, but while the orbits of the planets are reasonably circular those of most comets are very elongated. At its closest point to the Sun (perihelion) a comet is very close to the solar surface; at its furthest (aphelion) it may be far beyond Neptune and the Kuiper Belt. A comet is not a substantial, solid body like a planet; the head, usually only a few miles across, is composed of ice and rubble, sometimes likened to 'a dirty iceball'. Extending from it there may be a tail or tails, made up of dust particles or very tenuous gas. As a comet moves it leaves a dusty trail behind it; if one of these particles enters the Earth's atmosphere it will be heated by friction against the air particles and will burn up, producing a shooting star or meteor. A meteor will burn away at around forty miles above sea level, but a larger body may survive to reach the ground, and is then called a meteorite. Note that meteorites come from the asteroid belt, and are not associated with either comets or shooting star meteors.

The Sun is a star, shining by its own power; the surface is hot (over 5,000 degrees celsius) and the core has a temperature of about fifteen million degrees. The Sun is not burning in the manner of a coal fire; its energy is due to nuclear reactions going on deep inside it. It has been said that it is a vast, controlled atom bomb! It is indeed vast when compared with our world; you could cram a million Earths inside the Sun and still have room to spare. It is ninety-three million miles away, but in our sky it looks the same size as the Moon, which is much smaller than the Earth but is only about a quarter of a million miles from us.

Every star is a sun, shining by its own light. Some stars are less powerful than the Sun but we also know of stars which have well over a million times the Sun's luminosity. They look so much smaller and fainter than the Sun only because they are so much further away; the nearest star beyond the Sun is roughly twenty-four million million miles away. For distances of this kind, units such as the mile or the kilometre are inconveniently short (just as it would be clumsy to give the distance between London and Manchester in inches) and a different unit is preferable. Light does not travel instantaneously; it flashes along at 186,000 miles per second, so that in one year it crosses almost six million million miles. This distance is known as the light-year. The nearest star beyond the Sun is just over four light-years away.

The stars are so far away that their individual or proper motions are too slight to be noticed except over very long periods; the star patterns or constellations look virtually the same now as they must have done in the time of the Trojan War – it is only our near neighbours, the members of the Solar System, that move perceptibly from night to night. The constellations have been given attractive names, many of them mythological – Orion, Cassiopeia, Perseus and so on – but the stars in a constellation lie at different distances from us, and have no real connection with each other; we are dealing with nothing more than a line of sight, and the names mean nothing at all. We use the old Greek system, but the ancient Chinese and Egyptians used different constellation patterns and names.

The Sun is one of about 100,000 million suns making up our star system or Galaxy. Many of the stars have planets of their own, though as yet we have been unable to see them directly (except in a couple of rather dubious cases), and have had to locate them by indirect methods. The Galaxy is a flattened system, measuring 100,000 light-years from one side to the other; the Sun lies near the main plane, about 26,000 light-years from the centre. When we look along the main plane we see many stars in the same direction, and this causes the appearance of the Milky Way. The stars in the Milky Way are not really crowded together, and are in no danger of colliding; we are merely dealing with another line of sight effect.

As well as its individual stars, the Galaxy contains huge clouds of gas and dust called nebulae, inside which new stars are being formed from interstellar material. If a nebula is illuminated by a convenient star, it shines; if not, it is a dark mass detectable because it blocks the light from objects beyond it.

Our Galaxy is not the only one; we can see others – millions, hundreds of millions or even thousands of millions of light years from us. Galaxies tend to form groups or clusters; the Sun is a member of one such group (the Local Group). Each group of galaxies is receding from each other group, so that the entire universe is expanding – and the faster away they are, the faster they are receding. With modern instruments we can probe out to more than thirteen thousand million light-years.

It is now believed that the universe came into being 13.7 thousand million years ago; this is known (misleadingly) as the Big Bang theory, but we have to admit that we are reduced to little more than speculation.

Planets, satellites, stars, nebulae, galaxies ... This is a very rough outline of the make-up of the universe, but I hope that it is sufficient for the moment. Now let us begin our questions and answers.

If I want to have an astronomical telescope, could I make one?

You certainly could, and a few years ago, telescope making was very popular.

Telescopes, as you know, are of two main types: refractors and reflectors. A refractor collects its light by means of a lens known as an object glass, while a reflector uses a mirror. Making an object glass is really a task for the professional, but making a mirror is much easier, so that almost all home-made telescopes are reflectors. Most of these are Newtonian, because the optical system was first worked out by Sir Isaac Newton, who demonstrated his original telescope to the Royal Society in 1671. It had a mirror one inch across, but modern amateurs have made mirrors a great deal larger than this – up to several feet across.

In a Newtonian, the light from the target object passes down an open tube, and hits the main mirror (the speculum) at the lower end. The speculum is curved, and sends the light back up the tube on to a smaller flat mirror, placed at an angle of forty-five degrees. This flat mirror directs the rays into the side of the tube, where they are brought to focus and the image is enlarged by an eyepiece, which is essentially a magnifying glass. In a Newtonian, the observer looks into the tube rather than up it. The heart of the telescope is the speculum, which can be spherical but is much more effective if paraboloidal.

The trick here is to take two glass 'blanks' and rub one against the other, so that one becomes convex and the other

– destined to be the mirror – is concave. There is a special way of doing this; it takes a long time, and there are any number of things that can go wrong, but with sufficient patience it can be done. Most newcomers begin with six inch blanks; the flat and eyepiece can be bought at reasonable cost (actually you will want three eyepieces, one low powered, one medium, and one high). The rest of the telescope can be made by anyone who is reasonably 'handy'; there need not even be a solid tube, and many reflectors are skeletons. After all, the only function of the tube is to hold the optical components in the right positions.

Until very recently telescope making remained popular, because to buy even a reflector cost a great deal of money (and good refractors are always more expensive still). I used to advise against buying a reflector with a mirror less than six inches across, or a refractor with an object-glass with a diameter less than four inches – and a really useful telescope meant spending at least £300. The situation has changed; prices have come down, and it is possible to buy a small but adequate telescope for under £100. Of course it will be limited, but it will be much better than nothing at all, and home-made telescopes are becoming rather rare. Try your hand by all means, but be prepared for problems ...

Incidentally, do not despise binoculars. They cannot provide high magnification, but for some branches of observation they are surprisingly useful.

How far away is the Moon?
Is it the nearest body in the sky?

On average, the Moon is 238,000 miles away – rather less than a quarter of a million miles. This is by far the nearest natural celestial object, though of course we have launched many artificial satellites which are much closer. But the Moon's orbit is not a perfect circle; it is an ellipse, and the distance from us ranges between 252,000 miles and only 223,000 miles. At its closest it is said to be at perigee, and at its furthest it is at apogee.

The Moon is the only natural body which moves round the Earth. To be accurate, the Earth and Moon move together round their common centre of gravity, or barycentre, much as the bells of a dumbbell will do when you twist them by the bar joining them. However, the Earth is eighty-one times as massive as the Moon, and the barycentre lies deep inside the Earth's globe, so that the simple statement that 'the Moon goes round the Earth' is good enough for most purposes. The Moon takes 27.3 days to make one full circuit.

What is a Syzygy, and where can I find one?

You can't! This is the name given to the position of the Moon when new or full, so that the Earth, the Sun and the Moon are then lined up. Hideous word – it is pronounced 'sizzer-ji'.

Was the Moon ever part of the Earth?

Quite probably, but nobody is really sure how the Moon was formed, and all sorts of theories have been proposed.

We do have some facts to guide us; for example, we know that the Moon and the Earth are the same age – roughly 4.6 thousand million years. The Moon's mean density is lower than that of the Earth; the surface rocks are of the same general type, but the Moon has a smaller, heavy iron-rich core (remember that the Moon's diameter is only a little more than one quarter of the Earth's). However, the Moon is at least comparable with the Earth, and it is often said that the Earth-Moon system should be regarded as a double planet rather than as a planet and a satellite.

We are confident that the Earth is built up from the material of the 'solar nebula', a cloud of dust and gas surrounding the youthful Sun. It seems reasonable to think that the Moon condensed in the same way, at the same time and in the same region of the nebula, and this idea still has wide support, but there are various mathematical objections to it, because it would require a very special set of circumstances. Moreover it is not easy to explain the marked difference in density between the two globes. Alternatively, could the Moon have been formed in a different part of the nebula and later captured by the gravitational pull of the Earth? Again this sounds reasonable, but the mathematical difficulties are even greater.

A completely different scenario was given by George Darwin (son of the great naturalist Charles Darwin) in the latter part of the nineteenth century. Darwin pictured a combined Earth-Moon body which condensed from the nebula, and was initially hot and viscous. It was rotating, as do all bodies, but the spin was so rapid that the mass became unstable. Part of it was thrown off to build up the Moon, while the larger remaining part became the Earth. Again there seemed no obvious objections, and Darwin's theory was accepted for many years, but it simply does not work. A huge portion of material could not be hurled off in this way – and even if it could, there is no chance that a globe such as the Moon would be the result.

Today many astronomers – perhaps most – favour what is called the 'giant impact' theory. The original Earth-Moon body was hit by a massive impactor, perhaps almost the size of Mars. The cores of the two bodies merged, and *débris* was thrown around, but could not break completely free, so that after a comparatively short time it accreted to produce the Moon. At least this would account for the density difference, since the Moon would have built up from the outer, less substantial parts of the proto-Earth, and the theory seems to fit the facts better than the others, even though it does not solve all the problems.

En passant, it is worth recalling a comment made by Harold Urey, a Nobel laureate and one of the twentieth century's leading geophysists. According to Urey, because all theories of the Moon's origin are so unconvincing, science has proved that the Moon does not exist!

Are there many legends about the Moon?

Legends come from every country, and some of them are delightful. I particularly like a story which comes from China. A herd of elephants made a habit of drinking at the Moon Lake, and sometimes accidentally trampled upon the local hares. This would not do at all, but the chief hare, who was extremely clever, had the answer. He told the elephants that they were offending the Moon Goddess by disturbing her reflection in the water. The elephants agreed that this was most unwise, and made a hasty departure!

To the people of Van, in Turkey, the Moon was a young bachelor who was engaged to the Sun. Originally the Moon had shone in the daytime and the Sun at night, but the Sun, being a girl, was afraid of the dark, and so they changed places. (N.B. Please, no comments from politically-correct fanatics!)

Is there a dark side to the Moon?

Yes. Because the Moon is lit up by the Sun, one hemisphere is always bright, having its daytime, while the other hemisphere is dark, having its period of night. The Moon has a rotation period of 27.3 days (much longer than our own world's twenty-four hours!), so that a day or night on the Moon is equal to about a fortnight on Earth.

It is quite wrong to say that half of the Moon is permanently dark. It is true that the Moon always keeps the same face turned towards the Earth, but it does not always keep the same face towards the Sun, so that the day and night conditions are the same everywhere – except that on the 'far side' the nights would be darker, because the Earth would never be above the horizon.

What is Harvest Moon, and do other full moons have names?

In the northern hemisphere, the full moon closest to the autumnal equinox (around 22 September) is called Harvest Moon because the ecliptic then makes its shallowest angle with the horizon, and the retardation – the time-lapse between moonrise on successive nights – is at its minimum, and may be no more than fifteen minutes, although for most of the year it is closer to half an hour. The diagram illustrates what is meant; remember that from night to night the Moon covers the same distance along the ecliptic. It was held that this was useful for farmers gathering in their crops. A Harvest Moon

A Harvest Moon

looks the same as any other full moon – and certainly does not look larger than usual. The next full moon is known as Hunter's Moon. Other full moon names are seldom used. They are:

January	Winter Moon, Wolf Moon
February	Snow Moon, Hunger Moon
March	Lantern Moon, Crow Moon
April	Egg Moon, Planter's Moon
May	Flower Moon, Milk Moon
June	Rose Moon, Strawberry Moon
July	Thunder Moon, Hay Moon
August	Grain Moon, Green Corn Moon
September	Harvest Moon, Fruit Moon
October	Hunter's Moon, Falling Leaves Moon
November	Frosty Moon, Freezing Moon
December	Christmas Moon, Long Night Moon

What is the meaning behind the expression 'once in a blue moon' and what exactly is a blue moon?

The expression means 'something that happens only very occasionally', but astronomically there are two meanings. If there are two full moons in a calendar month, the second is said to be a blue moon – as for instance in January 1999, when the Moon was full on the 2nd and again on the 31st. This is not particularly uncommon, and does not indicate any unusual colour; it is not an old tradition, and comes from misinterpretation of comments made in an American periodical, the *Maine Farmers' Almanac*, in 1937. Nobody seems to know why these comments became so famous!

Yet the Moon can occasionally look blue due to conditions in the Earth's atmosphere; of course, all moonlight reaching us has first to pass through our air. For example, I well remember what happened on 26 September 1950, because of dust in the upper air raised by vast forest fires in Canada. I recorded that the Moon shone down in a lovely electric blue colour, unlike anything I have ever seen before. I did not see the blue moon of 27 August 1883, due to material sent up from the violent volcanic eruption at Krakatoa, but I have spoken to people who did see it, and apparently it was both eerie and spectacular. Green moons have been seen in Sweden in 1884, on 14 February in Kalmar and 12 January in Stockholm, though only for a few minutes each. No doubt there are many other instances of coloured full moons, but all these are mere atmospheric effects. Of course, the very low-down Moon often appears red (and so does the Sun).

I always think that a full moon looks almost as bright as the Sun. Is this true?

No, it most certainly is not! It would take roughly half a million full moons to equal the brilliance of the Sun. The Moon has no light of its own; it shines only by light reflected by the Sun. It is not even a good reflector. The surface rocks are surprisingly dark; the average reflecting power or albedo is less than ten per cent. Moreover, the Moon sends us very little heat, and this is why it is quite safe to look at it with a telescope or binoculars – whereas it is desperately dangerous to look at the Sun through any optical instrument (see pp. 126–128). With the full moon you may dazzle yourself, but nothing more, provided that you are sensible about it. If you stare for too long you will make your eyes extremely tired, which is not to be recommended.

Is our weather affected by the Moon?

Not directly. But of course the Moon is the main controller of our tides, and they do have an effect. Attempts to link the general weather with the Moon's phases have been, at best, inconclusive. There is for instance no evidence that the weather is better or worse at full moon than it is at any other time.

My home at Selsey, in Sussex, is a few hundred yards from the sea. Over the past half century I have tried to find a correlation between the weather at my meteorological station (2653 0007) and lunar phases, but once I had eliminated effects due purely to the tides I had no luck at all. Other investigators may well do better.

Mind you, Selsey and Bognor Regis have a mini-climate, with more sunshine, more clear skies and less cloud and rain than most areas – which is why King George V came to Bognor to recuperate, honoured it as Bognor Regis, and make that unfortunate remark about it!

I have sometimes seen a bright ring around the Moon, some distance from it. What is this and does the Moon ever cause rainbows?

A ring of this kind – known as a halo – is due to a thin layer of cloud in the Earth's air, called cirrostratus cloud, and lying at an altitude of at least 20,000 feet. You cannot see it, but the moonlight shining through it produces the halo.

Lunar rainbows do occur, but are much rarer than ordinary solar rainbows, because the Moon's light is so much weaker than that of the Sun, and there are no vivid colours. I have seen only one really good lunar rainbow; this was in 1942, when I was in an aircraft flying at about 8,000 feet above Scotland. Unfortunately I had no chance to pay much attention to it (I was the navigator of a bomber aircraft, returning from a raid over Germany) but I could see that the rainbow had a strange, ghostly beauty.

I do not know if a lunar rainbow has ever been photographed; if any of my readers has managed to secure such a picture, I would be most interested to see it.

I have heard that the phases of the Moon affect human behaviour. Is this true?

Scientifically it shouldn't be. It has been claimed that mentally disturbed people are at their worst at the time of full moon, but there is no reason why this should be so. The Moon's distance from the Earth does vary over the course of the month, because the lunar orbit is elliptical rather than circular, but the Moon's perigee (closest point) need not coincide with full phase, and may fall at any time in the month. Yet the belief persists, and it is quite true that many police patrols (those that are left!) tend to be alert on full-moon nights.

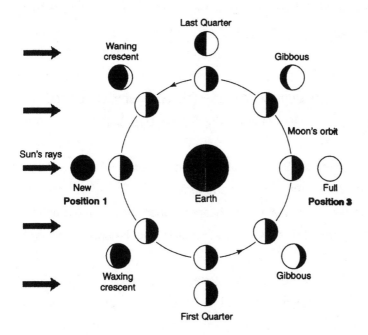

Phases of the Moon

I carried out a survey some time ago, and found that patrols and hospital workers tended to believe in a connection, while doctors were generally sceptical. As my own ignorance of medical matters is absolutely complete, there is no point in my making any comments.

There was one recent political episode. With the supreme confidence of the truly barmy, the Newcastle Green Party announced in 1992 that it would meet at new moon to discuss policies and ideas, and then at full moon to act upon them. They have not (yet) won any seats in Parliament, but one must wish them well. In 2003 I scanned some pages of *Hansard* to see if the speeches were any crazier at the time of full moon, but the standard of debate was so abysmally low at all lunar phases that I could come to no positive conclusions.

Is an eclipse of the Moon very different from an eclipse of the Sun?

Yes, quite different. A solar eclipse is caused when the Moon passes in front of the Sun and temporarily hides it, either totally or partially. A lunar eclipse happens when the Moon passes into the cone of shadow cast by the Earth, and its supply of sunlight is cut off.

As the Moon shines by reflecting the light of the Sun, you might expect it to disappear, but some of the Sun's rays are bent (refracted) on to the Moon by way of the layer of atmosphere surrounding the Earth. The Moon becomes dim,

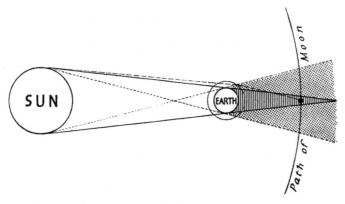

Theory of a lunar eclipse

often coppery coloured, and there may be lovely hues, so that an eclipse is always worth watching. Lunar eclipses may be either total or partial. They are leisurely affairs; totality may last for over an hour.

From any particular site on Earth, eclipses of the Moon are more frequent than those of the Sun. This is because when a lunar eclipse occurs it can be seen from any point on the Earth's surface where the Moon is above the horizon, whereas a solar eclipse is visible over only a restricted area.

In 1504 Christopher Columbus made use of a lunar eclipse. His ship was anchored off Jamaica, and the natives refused to send him supplies of food and water. Knowing that an eclipse was due, Columbus threatened to make the Moon 'change her colour and lose her light'. When the eclipse started, the Jamaicans were so terrified that they at once elevated Columbus to the status of a god, and provided him with all the supplies he needed!

Why has the Moon no air – and did it ever have any?

In the days when there was violent volcanic activity on the Moon, there must have been atmosphere, though of course not the kind of air we could breathe. However, atmosphere – whatever its composition – is made up of particles flying around at high speeds, and the Moon has a relatively weak gravitational pull, so that a particle moving out at a velocity of 1.5 miles per second must break free. The flying particles could move more quickly than this, so that the Moon could not retain them, and the lunar atmosphere leaked away into space until there was virtually none of it left. Not so with Earth, where the 'escape velocity' is seven miles per second rather than 1.5; our world could hold on to its air – or rather, the life gases in it. It seems certain that the Moon lost almost all its atmosphere before life had time to appear there, so that our satellite has always been sterile.

Is the Moon hot inside?

Yes. Many astronomers used to think that the Moon was cold and dead all the way though its globe, but the space missions have shown that this is not so. The Moon has a molten core, rich in iron, but it is much smaller than that of the Earth, both relatively and absolutely. It is probably about 600 miles in diameter, with a temperature of perhaps 1,500 degrees celsius.

The Earth and the Moon are of the same age, but the Moon cooled down much more quickly than the Earth, because it is so much smaller and less massive (a small Christmas pudding will cool down much more quickly than a large Christmas pudding!).Round the core comes the rocky mantle, and then the relatively thin surface layer, the top of which – the loose regolith – is nowhere more than a few tens of feet deep.

Before the space missions, there were astronomers who thought that the upper surface was covered with soft, deep dust-drifts, and Dr Thomas Gold of Cambridge even wrote that a spacecraft landing there would simply 'sink into the dust with all its gear'. These dire predictions were not finally disproved until 1966, when the Russian probe Luna 9 made a controlled landing in the Oceanus Procellarum (Ocean of Storms) and showed no sign of disappearing. Of course there may well be dangerous dust-drifts here and there, but so far, fortunately, we have not encountered them.

How were the Moon's craters formed?

Many astronomers used to think that they were volcanic (I did!) but we now know that they were the result of a cosmic bombardment which virtually ended over 2,500 million years ago. Huge pieces of matter – meteoroids – rained down on the lunar surface, and left their scars everywhere. Some of the craters are over 200 miles in diameter, while others are tiny pits too small to be seen from Earth.

Magnificent lunar craters. The sixty-four-mile Theophilus breaks into its neighbour
Cyrillus. Photograph by Bruce Kingsley, eleven-inch reflector

No part of the Moon is free from them; they are found on the bright areas, on the darker, smoother plains still called 'seas', and on both the Earth-facing and the far side of the Moon. A crater is essentially circular; when a meteoroid lands, it buries down below the lunar surface, and acts as a violent explosive, so that the result is a circular crater even if the impactor lands at an oblique angle. But, of course, older craters are broken and distorted by later impacts, and some are so ruined that they are almost unrecognisable. Generally, smaller craters break into larger ones – there are very few exceptions to this rule. Moreover, some craters have been overwhelmed by the lava flowing in the 'seas' while the Moon was still active; these old formations can be traced as 'ghost' craters. There are also craters whose seaward walls have been flooded over, so that they have been turned into bays.

Many craters have central mountains or groups of mountains, due to rebound after the initial impact. However, these central mountains never equal the heights of the crater walls – so that in theory it would be possible to put a lid over a crater without hitting the mountain-top! Note, too, that crater walls usually rise to only a modest height above the surrounding landscape; the floors are sunken. In shape, a crater is more like a shallow saucer than a steep-sided well or mineshaft. If you stand near the middle of the floor of a large crater (as no doubt astronauts will do, before long) you will not be able to see the outer walls at all; they will be below your horizon.

When you look at the Moon through a telescope, you will see that away from the centre of the disk the craters look elliptical (egg-shaped) rather than circular. This is because they are foreshortened; the Moon is a globe, not a flat plane. Seen from above, from a spacecraft, a large crater will be found to be circular, unless of course it has been broken and distorted by later impacts. Consider the dark-floored crater which we call Plato; it is sixty miles in diameter, and perfectly circular, but from Earth it shows up as an ellipse. Seen from Earth, the craters near the edge (limb) of the disk are so foreshortened that it is often almost impossible to distinguish a crater from a ridge. Before the space age, these parts of the Moon were very poorly mapped.

Who named the Moon's craters?

The first names, which we still use, were given by the Italian astronomer Giovanni Riccioli, who drew a Moon map in 1651. His system was to name the craters after famous men and women, usually, not always, astronomers. The seas were given romantic names such as the Sea of Tranquillity and the Bay of Rainbows; we always use these names in their Latin form, because Latin is still the universal language of science even though nobody has actually used it in conversation for a very long time. Thus the Sea of Tranquillity becomes Mare Tranquillitatis, the Bay of Rainbows is Sinus Iridum, and so on (*Mare* is Latin for 'sea'; plural *Maria*). Major mountain ranges are named after Earth ranges, such as the Alps and the Apennines.

The waning half moon. The Apennines are well shown. Photograph by Bruce Kingsley

The system is satisfactory enough, which is why we still use it, but it is far from perfect. Riccioli had his own ideas; he did not believe the Copernican theory that the Earth moves round the Sun, so that when he named a crater after Copernicus he 'flung it into the Ocean of Storm' (Oceanus Procellarum). He gave very important craters to himself and to his colleague Grimaldi, which was no doubt understandable! Later astronomers added other names, not always suitable; for example Newton has been allotted a very foreshortened walled plain, while Galileo is honoured only by a very obscure crater. Today the names are allotted by the appropriate Commission of the International Astronomical Union. Of course, a whole set of new names had to be introduced when we obtained the first photographs of the Moon's far side, which is never seen from Earth.

There was one amusing mistake. The first images of the far side were sent back by the USSR spacecraft *Lunik 3*, and the Russians instantly proposed new names. One feature seemed to be a major mountain range, and these peaks were at once christened the Soviet Mountains. Alas, later images showed that the feature was nothing more than a bright streak, and the Soviet Mountains were hastily deleted from official lunar maps!

If there are so many craters on the Moon, why are there not so many on the Earth?

Both the Earth and the Moon suffered the same cosmical bombardment in the remote past, and both were cratered.

But on the Moon there is no erosion – no air, no wind, no water to wear away the craters, so that they have been left undamaged. On Earth, the craters have been eroded away. We know some impact craters, notably Wolf Creek in Australia and the Barringer Crater in Arizona, but these are small and recent – that is to say they date back a few tens of thousands of years, not many millions.

Is it true that you will find Hell on the Moon?

Quite true. It is the name of a crater twenty miles in diameter, at 33°S, 8°W. It is quite well formed, and has a low central peak, but it is not unusually deep! It has been named after the well-known Hungarian astronomer Maximilian Hell, who lived between 1720 and 1792.

There are other names which may be regarded as unexpected, but commemorate past astronomers. Examples are Barrow (English astronomer Isaac Barrow, 1630–1677) and Birmingham (Irish astronomer, 1829–1884). Not many living people are honoured in this way; the general rule is that to have a crater named after you, you must either have been within a hundred miles of the Moon, or else you must be dead. Naturally, there are craters named after the *Apollo 11* astronauts, but all these – Aldrin, Armstrong and Collins – are very small; all the more prominent craters have long since been taken.

The lunar crater Pythagoras. It is really a circular diameter (ninety miles) but it is near the limb and it is foreshortened into an ellipse. Photograph by Bruce Kingsley

Do volcanic eruptions happen on the Moon?

They certainly did, well over two thousand million years ago. The Moon was then an active world, and magma flowed out from below the surface, flooding low-lying areas and the floors of some of the large impact craters. This was when the lunar 'seas' were produced. The lava flows ended rather suddenly, and since then there has been very little surface activity. The Moon today is a calm place.

But though major eruptions belong to the remote past, a little mild activity still goes on. For many years enthusiastic lunar observers reported occasional glows and local obscurations which became known as Transient Lunar Phenomena (TLP). These reports were questioned, mainly because most of the observers concerned (not all) were amateurs, but conclusive proof of their reality has now been obtained. 'Moving patches' inside the large crater Langrenus were recorded photographically by the French astronomer Audouin Dollfus in 1992, and there were also observations from Apollo spacecraft. Quite apart from these (and a famous spectroscopic record of a red glow in the crater Alphonsus obtained by the Russian observer N. Kozyrev in 1958), there are so many reliable reports that they cannot possibly be dismissed – though it is only too easy to be misled by effects in our own atmosphere, and published catalogues contain some entries which are certainly spurious.

The cause seems to be the release of gases trapped below the surface; solar heating must be involved as the Sun rises over TLP-prone areas. Events are most common in regions rich in rills; the area of Aristarchus, the most brilliant crater on the Moon, is a noted site. Of course the disturbances are very mild, and will pose no threat to future manned activities on the surface. Surface tremors also occur, but are too negligible to register on the Richter scale used to classify terrestrial earthquakes.

Is it true that the first really good map of the Moon was drawn by a German banker?

Well, more or less true. Wilhelm Beer was a successful banker in Berlin (his brother was Meyerbeer, the famous composer).

The 145-mile crater Clavius. Note the chain of craters across the floor. Photograph by Bruce Kingsley

He was interested in astronomy, and at his home had a small observatory, equipped with a 3.75-inch refracting telescope. To learn more, he called in a young professional astronomer, Johann von Mädler. They joined forces, and between 1830 and 1838 they used Beer's telescope to draw up an amazingly accurate lunar map. In 1839 they published it together with a book containing a complete description of the Moon's surface. Considering that they used a small telescope, it was a truly remarkable achievement, and their map remained the best for many years. They also made the first really useful maps of the surface of Mars.

Neither of them did much more lunar work after 1840, when Mädler left Berlin to become Director of the Dorpat Observatory in Estonia; he retired in 1865 and spent his last years in Hanover, where he died in 1874. Beer had died in 1850. Looking back at their work today, one is lost in admiration; to students of the Moon, the achievements of 'Beer and Mädler' will never be forgotten.

Is it true that a lunar crater vanished?

No, but there has been a great deal of discussion about the case of a small feature, Linné on the Mare Serentatis (Sea of Serenity) which was said to have turned from a crater into a white patch some time between 1839 and 1866.

Linné lies at latitude 27.7° N, 11.8° E. It is only one and a half miles in diameter, with a depth of one-third of a mile, but it is easy enough to locate; it is surrounded by a bright nimbus, due to ejecta from the original impact. There are relatively few craters on the Mare. By far the largest of these is Bessel (lat. 21.8° N, long. 17.9° E), which is eleven miles across and is crossed by a bright ray which runs from north to south and gives every impression of being associated with the Tycho system.

Linné was shown on Riccioli's map of 1651. It was also shown by the German observers Wilhelm Beer and Johann Mädler who produced the first good lunar map: their *Der Mond*, published in 1839, is a masterpiece of careful, accurate observation, and is amazingly good although carried out with a small telescope (Beer's 3.75-inch refractor). They described the crater as well formed, by far the most conspicuous feature in the area; there seemed to be nothing special about it.

35

Der Mond had one unexpected effect. It was widely believed that since the map by Beer and Mädler was so good, further observations of the Moon were unnecessary, and for several decades little attention was paid to lunar studies – except by one man, Julius Schmidt, Director of the Athens Observatory. In 1866 he made the spectacular announcement that the Linné Crater had disappeared, to be replaced by a mere white patch. This caused intense interest, and the attention of observers swung back to the Moon. What had happened? Had Linné really vanished?

Most people thought so. One of the most celebrated astronomers of the time, Angelo Secchi, used the powerful telescope at the Vatican Observatory and concluded that there was 'absolutely no doubt that a change had taken place'; another leading astronomer, Sir John Herschel, believed that a moonquake had shaken down the walls and filled the crater; a volcanic eruption seemed to be another possibility. But others questioned the reality of any change; in my view the last word was said by Mädler himself, who observed Linné in 1838 and again in 1867, concluding that there had been absolutely no alteration. Changes in the angle of illumination, even over short periods, show small features such as Linné in many different guises.

I have made many hundreds of observations of Linné using my own telescopes as well as the large refractors at the Lowell Observatory in Arizona and the Meudon Observatory in France. Generally it looks like a white patch with a tiny central spot, but if you catch it close to the terminator, and use a high power on an adequate telescope (my 12.5-inch reflector is very suitable) its true form is clear. Linné has played a rôle in the history of selenography, since after a long period of stagnation it forced observers to turn back to our neglected satellite.

The Moon was once a violently active place, with volcanic eruptions going on all the time and lava flowing out from below the crust. But things are very different now; the lunar world is calm, and major structural changes have long since ceased. The Moon looks much the same now as it must have done to the dinosaurs.

Could I ride a bicycle on the Moon?

Not very easily, I'm afraid. The problem is Moon dust. It is not deep (as some astronomers once believed it to be – I didn't), but it gets into everything, and clogs up all mechanical parts. It gets into spacesuits, clothes, and even astronauts' skin! How much of a hazard it will be to lunar cyclists we will not know until somebody actually tries it.

If I go to the Moon, will I find it easy to climb the lunar mountains?

Yes, because you will weigh so much less than you do on Earth, and the Moon's gravity is much less powerful than ours. If you weigh twelve stones on Earth, you will weigh only two stones on the Moon, which will make climbing easy. Against this, you will have to wear a spacesuit, and take the greatest care not to damage it.

You will also be helped by the fact that the mountains and the crater walls are not steep. The gradients are gentle everywhere, though it would not be a good idea to tumble into one of the major rills! There is no reason to believe that the mountains and walls will be unstable, and a falling rock will drop at a much lower speed, giving you ample time to dodge.

There will be various 'tourist attractions'. For example there is the Straight Wall, not far from the crater Thebit, which is not straight and is not a wall; it is a surface fault, sixty miles long. The ground drops away from east to west, and this shows up in a telescope – before full moon we see a black line, because the shadow of the fault is being cast on to the surface, while after full moon the line is bright, because the sunlight is falling on to the inclined surface of the fault. It was once thought that we really were dealing with a sheer cliff but in fact the gradient is so gentle that a tourist will be able to walk up it even when encumbered with space clothing.

Incidentally, it will be fascinating to visit the sites where the pioneer probes and astronauts landed. Most of the articles left on the Moon will doubtless be taken into lunar museums, but the footprints of Armstrong, Aldrin and others will still be visible; there is no weather on the Moon, and it will take a long time for the footprints to be covered up by meteoritic dust.

Is the Moon's far side much the same as the side we see from Earth?

Well, there are mountains, craters and valleys, but the two hemispheres are not identical, because the Moon has kept the same face turned toward us ever since an early stage in the history of the Earth-Moon system. Only one really large 'sea' lies mainly on the far side; this is the Mare Orientale or Eastern Sea, a vast ringed structure thought to be the youngest of the lunar maria. Only a tiny part of it can ever be seen from Earth, and then only under the best possible conditions. I first drew it in 1948, and thought that it was a new discovery; much later I learned that it had been seen forty years earlier by a German astronomer.

There are no other really large maria on the far side, and the crater arrangement is not quite the same as it is on the familiar hemisphere, but at least we know that there are no fundamental differences. Earlier there had been some very curious ideas; in the nineteenth century it was even suggested that the Moon's centre of gravity was lop-sided, and that all the air and water had been drawn round to the far side, which might well be inhabited. This would certainly have been a revelation, but we are now entitled to be confident that no living thing had ever existed on the Moon before Neil Armstrong and Buzz Aldrin arrived there in 1969.

Is it true that a man once tried to reach the Moon by using rockets fixed to his body?

Just possibly! There is a famous story that around 1400 a Chinese man named Wan Hoo strapped forty-seven rockets on to his body and told his servants to light them all at once. They did – and Wan Hoo vanished in a cloud of smoke, never to be seen again.

This story has been told time and time again. I have tried to track down its source, but with a total lack of success, and I am bound to say that I doubt whether there is any truth in it. If it really did happen, Wan Hoo must be regarded as the first victim of the science of astronautics.

How many men have walked on the Moon?

Up to the present time (2008) twelve – two each from *Apollos 11, 12, 14, 15, 16* and *17*.

Could I grow marigolds on the Moon?

I'm not quite sure! Until very recently I would have said 'No', but some experiments carried out in Ukraine in 2008 make me much less certain.

Scientists in Kiev planted marigolds in crushed anorthosite, which is an igneous rock very common on the lunar surface. At first there was no success, but when different kinds of bacteria were added the results were much more encouraging, and the marigolds started to survive remarkably well. Dr Bernard Foing, who worked with the Ukranian team, went on record as saying that an unmanned mission to the Moon could be equipped with tools that could crush the lunar rock before adding bacteria and seeds, so that 'the plants could form the basis of a precursor lunar ecosystem'. All this sounds intriguing – but remember! – we cannot completely reproduce the lunar environment. We can provide the anorthosite, reduce the atmospheric pressure almost to nil, and control the temperature, but we cannot easily simulate the lower gravity, only one-sixth as strong as that of the Earth. Moreover, the lunar surface is exposed to all sorts of radiations coming from space, while on Earth we are protected by our atmosphere.

So will lunar colonists of the future be able to gaze admiringly at beds of flowering marigolds? And if so, could this lead on to crops of all kinds? One would like to think so – but don't bank on it!

Will there ever be cities on the Moon?

I am quite sure that the answer is 'yes', always provided that we do not plunge into a world war which might even end in the destruction of civilisation. If we are to extend our activities into space, including colonisation first of the Moon and then of Mars, it is vital for all nations to work together. The days when Russia and America had the field more or less to themselves are long over.

All being well, there is no reason why a lunar base should not be in place by, say, the year 2030. Making forecasts is always dangerous, and 2030 may sound over-optimistic, but at the time when I write these words (2008) it seems reasonable enough. It may well take the form of a dome (better planned, we hope, than the farcical Millennium Dome!) and will have many uses; all sciences will benefit, because the Moon will be in many ways ideal for research. One of the earliest developments will be the setting up of a medical headquarters, also serving as a hospital for those people living in the base. Certainly there will be research centres of all kinds, including an astronomical observatory, plus a lunar equivalent of the Jet Propulsion Laboratory, Cape Canaveral and Baikonur. Quite apart from all this, there must be really comfortable living quarters and efficient kitchens. There is no reason why food on the Moon should be unpalatable!

Recreational facilities? These must come sooner rather than later, and again there seem to be no problems. No doubt outdoor activities will be very much to the fore, with guided tours of sites such as the Alpine Valley, Mount Pico and the Straight Wall.

Looking further ahead, we can visualise many separate centres – call them 'cities' if you like – perhaps linked by rail or by lunar 'motorways'. This may sound hopelessly futuristic today, but remember that when I began looking seriously at the Moon, around 1930, the whole concept of space travel was officially dismissed as due to the ravings of a few cranks. Even by the time of the outbreak of war, in 1939, members of organisations such as the British Interplanetary Society were regarded as amiable lunatics.

We have to realise that there may be unforeseen hazards which will put back these developments for many years. I have always believed radiation to be the most serious threat – there is no atmospheric shield on the Moon. Neither do we know the effects of living for a protracted period under conditions of one-sixth Earth gravity, and the problem may be unpleasantly obvious when the Moon-men return home, though of course it will be easy to make frequent journeys to and fro – something which will not be much more difficult for the denizens of Mars. We must take all the precautions we can.

There is one other point which is often overlooked. We know that one of the worst dangers we face today is that of over-population. The Earth is too crowded already, and in the relatively near future a real crisis will be upon us. It is tempting to claim that we can simply solve the problem by using the Moon, but this is utterly absurd, because living there must always be under very artificial conditions, and there can never be more than a few tens of thousands of residents at most. Trying to solve over-population in this way is rather like trying to cure traffic congestion by removing all the cars registered in Bognor Regis. It just doesn't work. We must think of something better.

So where does this leave us? Lunar bases must surely arrive but for some unforeseen scientific complication or a war triggered off by brainless political leaders. At first the bases will be purely scientific; then they will become social as well, and there will be real communities with their own special characteristics. At the beginning of the twentieth century the South Pole was out of reach; today there is a permanent observatory there, and it is easy to take a holiday cruise to Antarctica. The Mare Tranquillitatis is much more accessible to us than Antarctica was to my grandfather. You may go there yourself – who knows?

When we reach the Moon, will we able to fly around from one place to another?

Not by using ordinary aircraft, because these depend upon having air round them, and the lunar atmosphere is so thin that for all practical purposes you can forget it. Rockets will work, of course, because a rocket depends upon the principle of reaction, and operates best in a vacuum. No doubt lunar rocket planes will be built, but for a long time it seems that travel from one lunar base to another will have to be on the surface. Eventually there should be a proper railway network. One can only hope that it will be more efficient and more reliable than the British railways of the twenty-first century!

I am told that we owe our very existence to the Moon. Is this true?

It may well be. The Earth's axis is inclined to the perpendicular (to its orbit) at an angle of 23.5 degrees, and this is why we have our seasons. Even over very long periods of time this angle does not change much, and so the seasons stay almost the same. But it is the Moon which stabilises the Earth, and keeps the axial inclination practically constant. Without it, the angle would swing to and fro over periods of a few tens of thousands of years, and our climates would be wildly erratic – so erratic, in fact, that intelligent life might never have developed.

We can draw a comparison here with Mars, where the two moons (Phobos and Deimos) are much too small and lightweight to act as stabilisers. At present the axial tilt is almost the same as ours, but over a period of 100,000 years or so it oscillates between thirty-five degrees and only fourteen degrees. This would have caused problems for the Martians – if they had ever existed!

Does the Earth have a second Moon?

No – at least, not a second moon more than a few feet in diameter. If it existed, we would certainly have found it long ago. A small satellite reported in 1846 by F. Pettit, Director of the Toulouse Observatory, was certainly due to observational error, though the great French novelist Jules Verne found it very useful in his classic story, *Round the Moon* (1871) when the satellite perturbed the movement of the fictional projectile and sent it back to Earth!

Clouds of loose material reported in 1961 by the Polish astronomer K. Kordeylewski seem to exist, but patches of *débris* at stable points in the Earth's orbit are very different from a satellite. In 1898 Dr Georg Waltemath of Hamburg reported a whole flock of small moons, which certainly do not exist. In 1950 a very careful search for minor satellites was carried out by Clyde Tombaugh, discoverer of Pluto, with no success.

There are however some small bodies which move round the Sun, not round the Earth, but whose orbits are not very different from ours, though more eccentric. These bodies stay close to Earth over many orbits, and are known as quasi-satellites. The best known is asteroid 2753 Cruithne, three miles across, discovered in 1986 by Duncan Waidron with the UK Schmidt telescope at Siding Spring, Australia. With respect to the Earth its orbit is shaped rather like a kidney bean, but there is no danger of collision. Several similar bodies arthirty30 feet across, which shares our orbit round the Sun, first on one side of the Earth and then on the other. It never comes closer to us than 3.6 million miles. In 2660 and again in 3880 it will become a temporary Earth satellite, but it will not stay with us indefinitely.

Is the Moon moving away from the Earth?

Yes! The Moon is indeed moving away, due to tidal friction. This has been going on since a very early period in the story of the Earth-Moon system; the two bodies were once very close together (and remember, we are still not sure just how the Moon was formed). At present the Moon is receding at the rate of four centimetres per year, and this also means a tiny increase in the length of the Earth's rotation. On average, each day is 0.0000002 seconds longer than its predecessor, though this is not quite constant; there are slight irregularities due to movements deep inside the Earth.

This recession will not continue indefinitely. In theory it would cease when the length of the rotation period and the Moon's orbital period had increased to forty-seven times the present value, but this would take a very long time indeed – so long, in fact, that before then both Earth and Moon will have been destroyed by changes in the Sun. So you need be in no hurry to dash outdoors and look at the Moon before it disappears into the distance!

When do you think that we will set up the first proper lunar base?

It is now more than thirty years since the last men travelled to the Moon. Since then there have been many unmanned missions, and we now know a great deal more about our satellite than we did at the time of the Apollo missions; before long there is every reason to expect that we will go back. A lunar base may well be established by the middle of the present century.

The advantages are fairly obvious. For instance, there are medical experiments which can best be carried out under conditions of one-sixth Earth gravity, and a hospital will be high on the list of priorities; so will be a physical laboratory. Medical science will benefit in many ways, and there will undoubtedly be a lunar hospital. Working under conditions of one-sixth Earth gravity will also open up many new branches of research.

Astronomically, of course, the Moon is a near-perfect site, with no atmosphere to block incoming radiation, and the far side, shielded from Earth transmissions, is completely 'radio quiet'. All these advantages, and many others, make the Moon very attractive to scientific researchers. But a careful plan of campaign will have to be worked out, and this will involve a number of preparatory missions, both manned and unmanned, before the setting-up of a full-scale permanent base.

Two points of vital importance have to be taken into account. One is the danger of radiation, particularly at times of high solar activity – the Apollo astronauts seem to have been lucky in this respect. On the Moon there is no natural protection at all, and any major base will have to be provided with properly-made 'safe quarters' to be used whenever a solar storm is found to be imminent. Secondly, it is essential for all nations to work together. The space race between the USA and the USSR belongs to history, but if it is succeeded by conflict with China, whose space activities are now very much in the forefront of research, all plans for the peaceful exploration and subsequent colonisation of the Moon will be thrown into the melting pot.

When will we return to the Moon? Making forecasts is always dangerous, and all I can do here is to give my personal predictions, with the full knowledge that they may well be very wide of the mark. Given full co-operation, I estimate many brief missions during the next two decades; the start of setting up a major base around 2025, and permanent occupation by 2030 I wonder how right – or how wrong – I am. Only time will tell.

Who first saw Mercury, and who named it?

Mercury is never very conspicuous, because it is always in the same part of the sky as the Sun, and can never be seen against a dark background. The Greeks certainly knew it, but it was recorded much earlier by the Sumerians, who lived in the third millennium BC, but about whom we do not know a great deal. The Greeks named it Hermes after the messenger of the gods, because of its quick movement; the Latin name for Hermes is Mercury.

Mercury is closer to the Sun than we are. It is visible with the naked eye only in the west after sunset, or in the east before sunrise; before the fifth century BC it was thought that the 'evening star' and the 'morning star' were two different bodies (the same mistake was made with Venus). In fact Mercury is always elusive, so that you have to look for it when it is at its best. A word of warning here – if you want to sweep around looking for Mercury, use binoculars or a wide-field telescope by all means, but only when the Sun is completely below the horizon. Otherwise there is always the danger that you will look at the Sun, thereby damaging your eyes.

Mercury can generally be found if you wait patiently; from my observatory at Selsey I can generally find it for a couple of dozen evenings or mornings per year. Of course, I can find it whenever it is not too close to the Sun by using my telescope, because I can set the telescope to locate it even when it is invisible with the naked eye.

Can you ever see Mercury in the middle of the night?

No. Mercury always stays close to the Sun in the sky, so that when the Sun is well below the horizon, so is Mercury. You have to catch it for a fairly short period either after sunset or before sunrise.

If you have a properly-equipped telescope, you can of course see Mercury in the daytime, and this is the method used by astronomers who try to map the surface markings. But no Earth-based telescope will show much apart from the characteristic phase, and most of our knowledge comes from images sent back by space probes. Mercury is a desolate place; there are craters and valleys, and superficially the landscape looks very like that of the Moon, though there are significant differences in detail.

Mercury is closer to the Sun than Venus – but the surface of Venus is the hotter of the two. Why?

True; on average Venus is sixty-seven million miles from the Sun, Mercury only thirty-six. But Mercury has almost no atmosphere, while Venus has – and this is what makes the difference.

Venus' air is very thick – much denser than ours. It also consists mainly of the gas carbon dioxide, which acts like a greenhouse and shuts in the Sun's heat. Stand inside your greenhouse on a sunny day in summer, and you will soon become uncomfortably hot, because the glass is trapping the incoming solar rays. The carbon dioxide in the atmosphere of Venus acts in exactly this way, and the temperature soars to almost 1,000 degrees fahrenheit. Mercury has no such atmospheric 'blanket', though even so it is far too hot to support advanced life-forms of our type.

Could I live on Mercury?

Not in the open. There are several reasons for this. The first is that Mercury has, to all intents and purposes, no air. Air is made up of particles moving around at high speeds; if they move outward too quickly they will escape into space. The Earth pulls strongly enough to hold on to its air, but Mercury has a much weaker pull, so that it has lost any air it may once have had. A trace remains, but not enough to be of any use to us, so that if you are on Mercury you have to use breathing equipment (unless, of course, you are inside your spacecraft or in a protective base).

Mercury spins round very slowly – it takes over fifty-eight Earth days to make one rotation – and the temperatures range from intolerably hot to intolerably cold. Finally, it is on average a mere thirty-six million miles from the Sun, and receives the full blast of solar radiation, some of which is dangerous. All in all,

Mercury is not the place to go for a holiday, quite apart from the problem of getting there!

Is there any life on Mercury?

No life of the kind we can understand. Conditions there are hopelessty hostile, and we cannot imagine any living thing which could survive anywhere on Mercury. If there is any Mercurian life, it must be absolutely alien – and everything indicates that the planet has always been sterile.

When will I be able to go to Mercury?

Frankly, I cannot tell you. Men have been to the Moon and if all goes well we ought to reach Mars before the end of the present century, but Mercury is much more of a problem. There are several reasons for this. Mercury is much further away than Mars, and never comes much within fifty million miles of us; moreover, travelling in the inner part of the Solar System, relatively close to the Sun, involves extra risks. This is only the beginning; as we have noted (p. 51) Mercury is the very reverse of welcoming. There is no useful atmosphere, and over most of the surface the temperatures are intolerable.

Astronauts will either have to stay in their rocket, making only brief trips outside, or else set up a base – and this will be difficult, as bringing elaborate equipment on a trip of this kind is simply not practicable. It will be sensible to go underground, but we have no idea whether a suitable cave could be found. Remember, too, that it will be necessary to stay until Mercury and Earth are suitably placed for the return journey to start.

Whether we will ever establish a permanent base there seems doubtful, though an automatic observatory is more possible. I have absolutely no idea when astronauts will make the journey, but I am sure that it will not be yet awhile – if indeed it happens at all. So my answer to your question must be: 'Not yet, I'm afraid!'

If I could go to Mercury, would I be able to see the Earth?

Yes, whenever it was above the horizon. It would look like a bright, bluish star, and the Moon would also be visible. Nights on Mercury are very long – the axial rotation period is over fifty-seven Earth days – and the nights are also very dark, because there is almost no atmosphere, and the sky is clear all the time.

Is there a spider on Mercury?

Not if you mean a living spider! So far as we can tell, Mercury is a world utterly unsuited to any kind of life, at least life of the kind we can understand. The temperature conditions are not encouraging, and the atmosphere is absolutely negligible.

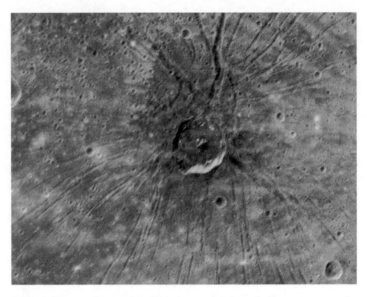

Crater on Mercury nicknamed the Spider. Image taken from the *Messenger* spacecraft

It is most improbable that living things ever appeared there, and Mercury has been sterile throughout its long history. But there is one very interesting crater which has been nicknamed the 'Spider'. It lies not far from the centre of Mercury's largest basin.

Almost all our knowledge about the surface features has been drawn from spacecraft, *Mariner 10* first and then *Messenger*. The best views to date are due to *Messenger*, which on 14 January 2008 swooped past the surface at a range of a mere 125 miles. The Caloris Basin proved to be 960 miles across, which is large when you remember that Mercury itself is only 3,030 miles in diameter. The floor of Caloris is brighter than the surrounding landscape, so that there is a great difference here between Mercury and the Moon; the lunar seas, such as the vast Mare Imbrium and the smaller Mare Crisium, are dark and depressed. And there, in the central part of Caloris, we find the twenty-five-mile Spider – a well-formed impact crater with a four-peaked mountain group not far from the centre. From it radiate dozens of troughs, and the overall pattern really does conjure up the impression of a web. Nothing like it has been found elsewhere on Mercury, and there is no lunar equivalent.

The Spider itself is certainly an impact crater, but the radiating grooves are mysterious. Presumably they were formed at the time when the meteorite landed, but even this is not certain, and we have to admit that we are still puzzled by this curious Mercurian arachnoid.

Can there be a planet closer to the Sun than Mercury?

There cannot be a large intra-Mercurian planet; we would certainly have found it by now. But little over a hundred years ago it was thought that such a planet did exist, and it was even given a name: Vulcan.

The Vulcan story began when astronomers found that Mercury was not moving quite in the way that it was expected to do. Something was wrong, and a French astronomer, Urbain Le Verrier, suspected that Mercury was being perturbed by a closer-in planet. If so, thought Le Verrier, it should be possible to track down the planet responsible. The main trouble was that if it were so close to the Sun it would be very difficult to see. The best chances would be either during a total eclipse of the Sun, when the sky becomes dark in the middle of the day, or when the planet passed in transit across the face of the Sun, as both Mercury and Venus do occasionally. (Incidentally, Le Verrier is said to have been one of the rudest men who has ever lived. One colleague commented that although he might not be the most detestable man in France, he was certainly the most detested!)

In 1859 Le Verrier was told that a transit of the planet had been observed by a French amateur astronomer, a Dr Lescarbault. Le Verrier went to see him, and found that he was not only the local doctor; he used to record his observations on planks of wood, planing them off when he had no further use for them; his telescope was small, and his timekeeper was a clock which had lost one of its hands! Yet amazingly, Le Verrier believed that he had really seen the planet, and worked out that it must be about twenty-one million miles from the Sun. It was then that the planet was named Vulcan, after the blacksmith of the gods. Le Verrier continued to believe in it for the rest of his life (he died in 1877).

However Vulcan was not seen again, either in transit or during total eclipses, and finally the irregularities in the motion of Mercury were cleared up by Einstein's theory of relativity. Vulcan, disappointingly, does not exist; it is one of the Solar System's ghosts.

On the other hand, there are many comets which invade these torrid regions. It has also been suggested that there may be in existence small bodies similar to asteroids; these are known as Vulcanoids – but so far, searches for them have been fruitless. If they are real, they will no doubt be found some time in the future.

Why was Venus given its name?

We can well understand why. The bright planets were named after the Olympian gods and goddesses. As seen with the naked eye, Venus is the loveliest of all the planets – so named in honour of the Goddess of Beauty. The ancients could not know that far from being a beautiful place, Venus is more like the conventional image of Hell. They would have been most surprised!

Why is Venus so bright, and can I see it in the daytime?

There are several reasons why Venus is much brighter than any other star or planet. First, it can come closer than any other large cosmic body apart from the Moon. Secondly, it is big – much larger than Mars, and very nearly as large as the Earth. Thirdly, its surface is permanently hidden by is thick, cloudy atmosphere – and clouds are very good at reflecting sunlight.

Keen-eyed people can see Venus in broad daylight, provided that they know just where to look. Because Venus, like Mercury, is closer to the Sun than we are, it always stays in the same part of the sky as the Sun, but at times it can rise several hours before sunrise or set several hours after the Sun has dropped below the horizon. It is then a magnificent sight, and can cast shadows.

People often think that Venus may have been the Star of Bethlehem. Do you think that this is true?

No, I don't. Remember, the Wise Men were astrologers, and they knew the sky very well. They were certainly familiar with Venus, and they could not possibly have mistaken it for anything unusual. If the Star of Bethlehem was really seen, it must have been something very unusual. No; Venus cannot be the answer, and neither can any other planet.

Is Venus called 'the Earth's twin'?

Some people do call it that, and in size and mass the two are very similar. Venus, slightly smaller and less massive, but the difference is really very slight. If I represent Earth and Venus by two snooker balls, they will be so alike that John Parrott and Stephen Hendry would quite happily use them to play a match!

If Venus is so like the Earth in size, why are conditions there so different?

The main cause must be that Venus is closer to the Sun: only sixty-seven million miles out, against ninety-three million miles for our world.

We are sure that Venus and the Earth were born at the same time, 4.6 thousand million years ago, and they probably started to evolve in the same way, with lands and seas. But at that time the Sun was not as powerful as it is now, and as it became older it also became more luminous. Earth was distant enough to be safe; Venus was not, so that as the temperatures rose the oceans of Venus boiled away, and the carbonates were driven out of the rocks on the surface. By now the atmosphere was rich in carbon dioxide, and Venus went through what we may call a 'runaway greenhouse' period. In a short time, cosmically speaking, it was transformed from a friendly world, suited to support life, into the furnace-like planet of today.

Did life ever begin there? We do not know. If it did, it cannot have developed very far before the conditions became hopelessly hostile, so that all life-forms were ruthlessly snuffed out. If astronauts go there, it is not impossible that they will find traces of primitive past life, but nothing more – and certainly no dinosaur skeletons! Personally, I doubt whether life ever gained a foothold there.

Can Venus ever pass in front of the Sun?

Indeed it can. When this happens, Venus is seen in transit as a black disk against the brilliant surface. At every 'inferior conjunction', when Venus is new and has its night side facing us, the planet is more or less between the Earth and the Sun, but its orbit is inclined to ours at an angle of about 3.5 degrees, so that at most inferior conjunctions Venus passes either above or below the Sun in the sky, and avoids transit.

Transits occur in pairs; one transit is followed by another eight years later after which there are no more for over a century. Thus there were transits in 1761, 1769, 1874 and 1882; the next, 2004 and 2012, after which we must wait until 2117 and 2125. During transit, Venus, unlike Mercury, is an easy naked-eye object, blacker than any sunspot.

Transits of Venus were once regarded as very important, because they provided a method of measuring the length of the astronomical unit (Earth-Sun distance). Observers had to time the exact moment when Venus passed on to the solar disk, and observers were ready to make long journeys to the most favourable sites. In 1769 Captain Cook was detailed to take the astronomer Charles Green to Tahiti, where the transit was expected to be well seen, and the observations were duly made. Cook then continued the voyage – and discovered Australia.

Unfortunately, the accuracy of the method was ruined by the presence of Venus' atmosphere. As the plane passes on to the Sun's face it seems to draw a strip of blackness after it, and when the strip disappears the transit is already in progress. (I saw this 'Black Drop' myself at the transit of 2004.) In any case the method is now obsolete, so that future transits will be regarded merely as interesting spectacles.

Does Venus have a moon like ours?

No. If Venus had a satellite of appreciable size, it would be easy to find. In the past, some observers using telescopes have claimed that they have seen a satellite, but they were certainly mistaken. Venus is so brilliant that it can produce 'ghost images', due to tiny imperfections in telescopic optics.

Careful searches have been made, and we are now confident that there can be no satellite more than a few inches across. Venus, like Mercury, is a solitary traveller in space.

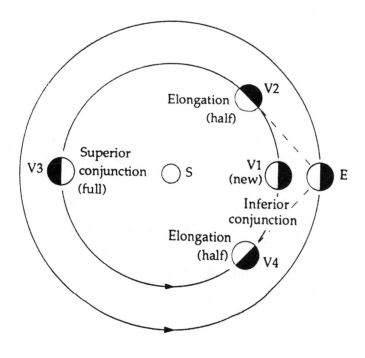

Phases of Venus

Isn't there a story about Napoleon and the planet Venus?

Yes, there is – and it is probably true!

It goes back about two hundred years. Between 1800 and 1815 France was ruled by Napoleon, who wanted to conquer the world – just as the German dictator, Adolf Hitler, planned to do in the 1930s. In a way the careers of these two leaders were similar. Each invaded and occupied country after country; each, for a while, controlled almost the whole of Europe – except Britain. In 1815 Napoleon was finally beaten at the Battle of Waterloo, in which our commander was the Duke of Wellington; in 1945 Germany surrendered to forces which now included Americans.

On one occasion the French held a great ceremony in honour of Napoleon. However, it became obvious that many people were staring at a particular spot in the sky instead of looking at Napoleon. Offended, the Emperor asked why he was not the centre of attention – after all, it was meant to be his parade! Someone explained that the object in the sky was 'a star'– in fact it was Venus. What Napoleon thought about it is not on record, but it is unlikely that he was pleased. No doubt the members of the audience were given strict and prompt orders to turn away from the sky and turn back to Napoleon.

Is Venus closer to us than Mars?

Not always, of course. All the planets including the Earth, move round the Sun, and when Venus is on the far side of the Sun – at the position we call 'superior conjunction' – it may be over 160 million miles away. But it is true than when closest to us, its distance may be no more than twenty-four million miles, and Mars can never be as close as that. The minimum distance between Mars and the Earth is still over thirty-four million miles. In fact Venus is the nearest large body in the sky apart from the Moon.

How long is a 'day' on Venus?

Venus has a most peculiar calendar. The 'year' – that is to say, the time taken by Venus to go once round the Sun – is equal to 225 Earth days (more precisely, 224.7), but the axial rotation period is 243 Earth days. In fact, on Venus the solar day is longer than the year. To make the situation even more curious, Venus spins in the opposite sense to the Earth, so that the Sun would rise in the west and set in the east; the interval between one sunrise and the next would be 118 Earth days. Mind you, you could never see the Sun properly from the surface, because the cloud-cover is total and permanent; there is no such thing as a sunny day, though the position of the Sun would presumably be indicated by an ill-defined glare.

We have to admit that we do not know why Venus spins in this extraordinary way. It has been widely believed that in the early life of the Solar System the globe was struck by a massive impactor and literally tipped over, but it may be more likely that interactions between the inner planets were responsible. We have solved many of Venus' mysteries, but this one continues to baffle us.

Are there lands and seas on Venus?

Venus is all 'land'; the surface is so hot that there can be no seas. Of course, water would boil away at once, and no other kinds of liquids could exist either. Venus is a world of volcanic plains, highlands and lowlands, and there are high mountains.

There are two main highland areas, named Ishtar Terra and Aphrodite Terra. Ishtar, in the northern hemisphere of the planet, is about the size of Australia, and is made up of two separate components separated by the Maxwell Mountains, the highest peaks on Venus. Aphrodite, mainly in the southern hemisphere, is larger, and is also made up of two masses with a lower area in between. It is not unlikely that the plains were once watery, but we cannot be sure; there may well have been seas after Venus solidified and developed a crust – but even if so, the increasing power of the Sun soon evaporated them. Today, lava-flows are everywhere, and from our point of view the conditions could hardly be more hostile.

We have to admit that in some ways Venus has been a disappointment. Around 1960 I remember giving a talk at Cambridge University and saying that as a potential colony

Venus might be a better bet than Mars. How wrong I was! But others also thought so, and Venus was the prime target of the early spacecraft, both Russian and American. When the real nature of Venus became known, the space-planners' main attention shifted back to Mars, and for a while it almost seemed that Venus was being neglected. Now, however, the exploration of the planet is once more under way.

Can I go for a walk on Venus?

It would be decidedly difficult, even if you could get there. At the surface of Venus the windspeeds are very low – but in that super-dense atmosphere they have tremendous force. A 5mph Venus breeze would be as forceful as a terrestrial hurricane, and I doubt whether any human walker could keep upright. No doubt somebody will try eventually, but not for a long time yet. Venus is not the place for a London-type marathon!

Could I climb the mountains of Venus?

Theoretically I suppose you could, but it wouldn't be very easy. You will have to wear full space equipment, the temperature will be forbiddingly high, and do not forget that you will probably have to contend with a drizzle of 'rain', not of water, but of sulphuric acid droplets.

If you intend to tackle the Maxwell Mountains, Venus' equivalent of our Himalayas, you will find slopes of up to thirty-five degrees, perhaps even steeper (quite unlike the gentle gradients of Martian volcanoes). Remember, too, that the surface of Venus is likely to be very unstable, and that you will weigh almost as much as you do at home.

Plan an expedition by all means, but please forgive me if I decide not to join you!

If I go to Venus, will I see dinosaurs?

No, I'm afraid not – but if you'd asked me that question a hundred years ago, I might have replied 'Possibly!'

We have known for a long time that Venus has a thick atmosphere, but before the space age we knew nothing about what we would find on its surface. Very reasonably, some astronomers believed that it might now be in a condition similar to that of the Earth hundreds of millions of years ago. Svante Arrhenius, of Sweden – whose work was good enough to win him a Nobel Prize – believed that Venus had warm swamps with luxuriant tropical vegetation; dragonflies flitted between the ferns, and there were amphibians starting to crawl out of the swamps. Others pictured Venus as being in a stage where reptiles had appeared, so that dinosaurs wandered around the hot, wet plains and swam in the shallow seas. If this had been so, astronauts would indeed have seen plesiosaurs, ichthyosaurs and pterodactyls – and would have had to watch for any sign of an approaching Tyrannosaurus rex!

But the first space missions ruled out anything of this kind. Instead of being no more than pleasantly warm, Venus is scorching hot, and no dinosaur could tolerate a temperature not far short of 1,000 degrees fahrenheit. Obviously there is no water, and not a trace of vegetation. So if we want to locate dinosaurs, we cannot look to Venus, or for that matter anywhere else in the Solar System. If dinosaurs exist – and I see no reason why not – they must live upon a planet of another star, light-years away.

Why is Mars red?

Because its surface really is red – or at least ochre. There are side areas which we call deserts, and this isn't a bad name for them, but they are not like our deserts such as the Sahara. Instead of yellow sand, they are covered with reddish mineral – 'rust'. This is what gives Mars the characteristic colour which made the ancients name it in honour of the God of War.

There is another difference, too. Our deserts are hot, at least in the daytime, but the Martian deserts are decidedly chilly, because Mars is much further away from the Sun than we are (on average 141.5 million miles, against ninety-three million miles for the Earth). At noon on a summer day it can be reasonably warm there, but long before nightfall the temperature has dropped well below freezing point, and the nights are colder than anywhere on Earth. This is true even at the equator of Mars.

Are there any seas on Mars?

Not now, but there used to be, several thousands of millions of years ago. We know that there is plenty of ice, not only at the poles but also elsewhere, and we can see unmistakable evidence of past water action. There are riverbeds, which are now dry once rushing torrents; we see what are obviously old islands, and we can tell positions of the old seas. In those days Mars was a warmer, wetter and friendlier world.

The seas lasted for a long time, but the gravitational pull of Mars is much weaker than that of the Earth, and the atmosphere leaked away into space, so that the seas dried up. It is possible that liquid water still exists not far below the surface, and springs may occasionally gush out; our spacecraft have given clear indications of this, but when water reaches the surface it quickly evaporates. It is not likely that there are extensive underground oceans, though of course we cannot be quite sure.

I have heard about the canals on Mars. Do they really exist?

I'm afraid not, but a hundred years ago many people thought that they did.

The story of the canals began in 1877, when the Italian astronomer Giovanni Schiaparelli was making drawings of Mars. He was a well-known observer, and he was using a good telescope (a nine-inch refractor) in Milan. He saw the white

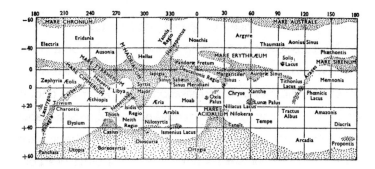

My map of Mars, drawn from observations with a 12.5-inch telescope

polar caps, which he believed (correctly) to be ice, and also the dark areas, which he believed (incorrectly) to be tracts of vegetation. But he also saw – or thought he saw – something else; thin, regular lines crossing the red 'deserts'. He called them *canali*. This is Italian for 'channels', but was inevitably translated as 'canals', and canals they remained. For some years after 1877 nobody else managed to see them, but before long other astronomers started to report them, and by 1890 canals had become all the rage. Schiaparelli did not pretend to know what they were, but then the suggestion was made they might be artificial waterways, built by the Martians to convey water from the icecaps through to the centres of population. This was certainly the view of Percival Lowell, who set up a major observatory at Flagstaff, in Arizona, and equipped it with a fine twenty-four-inch refractor. It was one of the best refracting telescopes in the world – and indeed, it still is. Between 1894 and 1916 Lowell and his assistants observed Mars on every possible occasion, and drew maps covering the planet with a network of canals. There was nothing haphazard about it; to

Lowell it was evidence of a global irrigation system, built by the Martians to use every scrap of water available on a world which was frighteningly dry. Lowell died in 1916, unshaken in his belief in a brilliant Martian civilisation. However, many people disagreed. Some astronomers could not see the canals even with telescopes as powerful as Lowell's, while others could make out only vague streaks, not in the least artificial in aspect. Certainly the canals could not be open water; if they existed, they would have to be narrow channels flanked by strips of vegetation. The question was finally settled by the post-war spacecraft. There are no canals; they were due simply to tricks of the eye. When trying to glimpse details at the very limit of visibility, it is only too easy to 'see' what you half expect to see.

I well remember my first view of Mars through Lowell's telescope, after the war. I had been at Flagstaff, busy with Moon-mapping; Mars was well placed, and after moonset I turned the twenty-four-inch refractor toward it. This, remember, was years before the start of the space age. Would I see canals? I am delighted to say that I didn't!

What will the Earth look like from Mars?

From the surface of Mars, the Earth will look like a bright star when conditions are favourable. It will be brighter than Mars appears to us, because the Earth is larger than Mars and is also better at reflecting sunlight. It will look blue, and the Moon will be an easy naked-eye object.

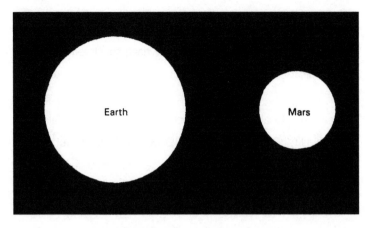

Comparative sizes of Earth and Mars

Earth will behave in the way that Venus does to us, and will remain in the same part of the sky as the Sun – in the west after sunset, or in the east before dawn. Of course, when Earth and Mars are on opposite sides of the Sun – that is to say, when we see Mars at opposition – Martian observers will be unable to see Earth for a brief period.

When there are manned observatories on Mars, details on the Earth – and the Moon – will be seen easily enough. Unmanned probes, such as *Mars Reconnaissance Orbiter* and *Mars Express*, have already (in 2007) sent back good pictures.

If I stand on Mars and shout, will I be heard?

If you are inside a Martian base, of course you will, but in the open it will be different. You will be wearing breathing equipment, and in any case the atmosphere is so thin that it will not carry sound waves well, so that your voice will seem very weak. In the open, conversations will have to be carried on by using radio.

Will my magnetic compass work on Mars?

No. The magnetic field of Mars is so weak that it could not affect the needle of your compass.

We know the reason for this. The Earth's outer core is liquid, and is rich in iron. This core is in rotation and swirling currents produce a magnetic field. But Mars no longer possesses a core like this, if indeed it ever had one. The core is much smaller than ours, both relativity and absolutely; we are not sure whether any part of it is liquid, and the general magnetic field is absolutely negligible. To make things even more difficult, there are localised magnetic areas here and there on the Martian surface, so that your compass might be actually misleading. Neither would your compass be of any use on Venus or the Moon.

Do Marsquakes occur – and if so, would they be strong enough to damage any Martian bases we set up?

Ground tremors do occur, and have been measured by the instruments taken to Mars by our spacecraft, but violent vulcanism belongs to the remote past, at least so far as we can tell.

Our future bases are not likely to be damaged in this way. By the time we are ready to build them we will know much more about Marsquakes than we do now, and will be able to make our bases firm enough to withstand the most powerful shock that can be encountered.

Could birds fly on Mars?

Birds of the kind you see in the garden certainly could not. The atmosphere is much thinner than ours, and a bird such as a robin could not produce enough 'lift' even by frantic flapping. But a bird with a tighter body and more effective wings might be able to manage. Given enough time, I can see no reason why Martian birds should not appear, but there are several things to be borne in mind. First, we are not sure if life really has appeared on Mars at any stage; however let us assume that it started there at the same time that it did on Earth. Terrestrial life developed steadily, though not quickly; there were primitive sea creatures, then fish, then amphibians, then reptiles such as the dinosaurs, then mammals – including birds. (Some members of the dinosaur family, notably the pterodactyls, took to the air, but they were probably not wing-flappers, and relied mainly upon gliding.) True birds date back only a few hundred million years, and by then Mars had become much less hospitable; the weaker gravity meant that much of the air had leaked away into space. In fact, Martian birds simply had insufficient time to evolve. (Dragonflies no doubt did, but a dragonfly is an insect.)

So — no birds on Mars. There never will be unless, in the future, we colonise the planet, make it suitable for life of our kind, and find some way to thicken the atmosphere and make it fit to breathe. This may happen eventually, but not for a very long time. If you holiday on Mars within the next few centuries, I fear that there is no hope of waking at sunrise to enjoy the twittering of a dawn chorus — for this, we must look to a world in another stellar system, and interstellar travel is beyond our reach as yet.

Does Mars have a pole star like ours?

It has an even brighter north pole star. We have the second-magnitude Polaris, in the constellation of Ursa Minor (the Little Bear), but Mars has Deneb in Cygnus (the Swan), which is a magnitude brighter than Polaris. Mars also has a more conspicuous south pole star. Ours is Sigma Octantis, in the Octant, which is none too easy to see with the naked eye even when the sky is clear, but Mars has the 2.5 magnitude Markeb or Kappa Velorum, in Vela (the Sails), which is particularly easy to identify because it is one of the stars making up the pattern of the False Cross (the three others are Koo She in Vela, and Avior and Tureis in Carina, the Keel).

From Mars, the constellation patterns would look the same as they do to us, because the distance between Mars and ourselves is so small compared with the distances of the stars, but from Mars a new, brilliant blue planet would often be seen — our own Earth!

If I am on Mars, in the open air, can I boil a kettle of water?

Only too easily! The boiling point of water depends on the atmospheric pressure. At sea level on Earth, water boils at 212 degrees fahrenheit. Go to the summit of Everest, where the air pressure is less, and water will boil at 187 degrees fahrenheit. The less the pressure is, the lower the boiling point.

Now go to Mars, and come out of your pressurised base. Outdoors, the pressure is below ten millibars – and this is too low for liquid water to exist. It would promptly boil away – in fact, evaporate. So if you want to make a cup of tea, go back inside your base!

Bear in mind, too, that blood is liquid. Go out on to the Martian surface, unprotected by a pressure suit, and your blood would literally boil inside you, with disastrous results. Short of thickening the atmosphere, there is nothing that we can do about this, and at the moment there is no way of turning Mars back into the warmer, wetter world it must once have been.

Could I go skating on Mars?

Mars is certainly cold enough, but there are no frozen seas. There may have been oceans many millions of years ago, but all the surface water has disappeared, so that skating would be difficult. Whether there are any areas icy enough to satisfy would-be skaters is very uncertain.

Of course, if we set up true Martian bases – cities, if you like – we will be able to make skating rinks, but outdoor activities mean wearing spacesuits, which will always be cumbersome, even though you will have only one-third your Earth weight.

Are there any polar bears on Mars?

There is plenty of ice at the Martian poles, and there are extensive, thick ice-caps – but as we know, the atmosphere there is quite unbreathable by any advanced creatures of the type we can understand. Quite apart from being much too thin, it has very little free oxygen, so that even the hardiest polar bear would be unable to survive. Future explorers will not have to face large, hungry white animals!

On Mars, do the polar ice-caps melt in the summer?

No, but they 'sublime' – that is to say, change straight from solid to gas without going through the liquid stage.

In Martian winter the polar caps can be very extensive. The caps are made mainly of water ice (with a seasonal coating of carbon dioxide ice), and when warmer weather arrives in spring they start to shrink, becoming very small by midsummer. It is fascinating to follow the shrinking of the caps, and under good conditions this is possible even with a small telescope. The caps sublime, rather then melt, because the atmospheric pressure is so low: less than ten millibars. As we have noted, liquid water could not persist on the surface; it would quickly evaporate.

I have heard that Bigfoot has been seen on Mars. Is this true?

Definitely not – but I can tell you where this remarkable rumour originated!

Bigfoot is a creature which some people believe to live in very inaccessible parts of the Earth. Its existence has never been proved, and I very much doubt its reality. In January 2008 some enthusiastic folk who were examining images sent back by NASA's Martian rover, *Spirit*, from inside the crater Gusev suddenly announced that they had found a feature which looked humanoid, and not unlike some of the vague impressions of Bigfoot. Needless to say, all the flying saucerers and astrologers were delighted to join in the fun, and produced yet again the famous 'face on Mars'. But what exactly was the *Spirit* image?

The answer is – a rock, standing at angle. This makes up the 'body'. Well behind it, lower down and quite separate, are two patches of darker rock accounting for the 'head' and the 'body'. Quite simple – in fact, disappointingly so, but no astrologer or UFO enthusiast will believe it.

Others compared the *Spirit* image with the Mermaid statue in Copenhagen. Take your pick. But in any event, I am quite sure that we have not heard the last of this particular image.

If we found living things on Mars, would it be safe to bring them to the Earth?

Almost certainly 'yes', but there would always be a worrying doubt until very full examination had been made.

Remember, at present we have absolutely no proof of life anywhere except on Earth, and if we do find life on Mars it could be very different from our own. Contact with it could be dangerous; we could be 'infected' in some way, and we would have no idea of how to deal with the situation. Also, there might be effects on plant life. We would be faced with a crisis unlike any other. Therefore, any Martian life would have to be studied for long enough to ensure that it was safe.

When the first astronauts came back from the Moon, they were quarantined until the examinations had been completed; quarantining was abandoned after *Apollo 12*, because by then we were satisfied that the Moon is sterile, and has always been so. But Mars does have atmosphere, and there is a good chance there was life there long ago; it may survive even now. I would say that the chance of danger is at least a thousand to one against, but we cannot be 100 per cent certain until laboratory tests have been made. Probably these will be carried out in space, or on a space.

Beyond Mars? Well, the other members of the Solar System do not seem to be menacing. If we make contact with life on planets of other stars, the whole situation is different – but at the moment this seems to be a long way ahead.

There are two other points worth making. If there were 'Martians', they might not react well to being brought to Earth, where the surface gravity is three times as strong. And just as we have to be wary of them, they would be entitled to be equally wary of us!

Would Martians be unfriendly?

I can give you an answer, but first there is one important point to be made. It is highly unlikely that there ever have been any Martians.

Mars and the Earth were almost certainly formed at the same time, from the solar nebula, about 4.6 thousand million years ago. Probably they started to evolve in the same way, but there was one vital difference. Mars is much less massive than the Earth, with a much weaker pull of gravity. Our escape velocity is seven miles per second, but that of Mars is only just over three miles per second. Therefore, Earth could hold on to its atmosphere, and life had time to develop. By the time this happened, Mars had lost much of its air, which had leaked away into space. This means that there had been no time for advanced life to evolve. Martian life was limited to very simple forms; there can never have been anything so advanced as an earwig.

This rules out intelligent Martians, but just for a moment suppose that we are wrong, and that there was once a true civilisation there. There is not the slightest reason to assume that it would have been hostile. Ideas of this kind go back to H.G. Wells, one of our greatest novelists. About a hundred years ago he wrote a story called *The War of the Worlds*, in which Earth is invaded by repulsive Martian monsters who are intent upon taking over our planet, and are finally destroyed only by terrestrial bacteria against which they have no protection. Just before the war, a clever adaptation of the novel was broadcast on American radio – and was taken to be a factual news bulletin. Thousands of people claimed that they had seen the machines, felt the heat-rays and watched helpless victims being butchered...It was some time before the panic subsided. True, it is most improbable that people in Britain would have been taken in, but it does stress the power of self-deception. (Remember the stories told by people who claim to have been abducted by extraterrestrials in flying saucers!)

The trouble was that many much less skilful writers followed Wells' lead, and it became customary to depict Martians as evil and inhuman. This is totally unjustified. Had Martians evolved, they might well have ended up by being much more civilised than we are. In only a few thousands of years, *Homo sapiens* has progressed from cultivating crops to making atom bombs which have already killed tens of thousands of innocent people, and in the near future may kill tens of thousands more. Martians might have done better; we have to admit that they could hardly have done worse! Personally, I have to say that I am sorry they never existed; we might have learned a great deal from them.

When was the first spacecraft sent to Mars?

The story really began in 1962, but the first success was delayed until 1965 – there is one curious fact; so far as Mars is concerned, the Russians have had no luck at all, despite their successes with Venus – which ought to be a much more difficult target. The Japanese have also failed, so that what we have learned is due entirely to America and, latterly, Europe.

Russia (then the USSR) started off with *Mars 1* (launched 1 November 1962). The actual take-off went well, but when the probe was about 70,000 miles from Earth, contact was lost. Probably Mars was passed in June at a range of 190,000 miles, but we cannot be sure, because nothing more was ever heard from it. It was not a promising beginning.

NASA was next in the field, and in November 1964 launched two instrument-carrying *Mariners 3* and *4*. *Mariner 1* and *2* were Venus probes. The first of these failed because in the pre-launch reparations someone forgot to feed a minus sign into a computer, but the second was a great success, and gave us our first reliable close-range information about that unfriendly world. *Mariner 3* also failed, because a shield designed to protect it during the first stages of the flight could not be dislodged, but its twin, *Mariner 4*, more than made up for this. On 14 July 1965 it flew past Mars at a range of just over 6,000 miles, and sent back twenty-one pictures, some of which were pleasingly clear. Only one per cent of the total surface of the planet was imaged, but the results showed that many of our ideas were wrong. We had expected a fairly level surface, with no high mountains or deep valleys. Instead, we found a cratered landscape, more reminiscent of the Moon than of the Earth. Canals were conspicuous only by their absence; it was clear that the dark areas were not coated with vegetation, and the cold was numbing. Perhaps most important of all, the atmosphere was much thinner than expected. The ground pressure was only around ten millibars, and the main constituent was presumably carbon dioxde rather than oxygen. Any advanced life-forms were finally ruled out; the question now was – did Martian life exist at all?

Mariner 4's work was done; no more data was received, even though it remained in touch until December 1967. No doubt it is still orbiting the Sun, but we have no hope of contacting it again. Following another Russian failure (*Zond 2*) we come to NASA's *Mainers 6* and 7, which by-passed Mars in the summer of 1969 and returned useful images: seventy-six from *Mariner 6*, and 126 from *Mariner 7*. More plains, more craters. Even at NASA, enthusiasm was on the wane. I was there in August of that year, and found that Mars was now being dismissed as boring. We know that by sheer bad luck, all the first fly-by probes passed over the least interesting areas of the planet.

However, NASA did not give up, and returned to the attack in 1971. *Mariner 8* was a prompt failure, because the second stage of the launcher failed to ignite, and the spacecraft made an undignified descent into the sea, but its twin followed less than a month later, and reached its target on 13 November. This was no fly-by mission. *Mariner 9* fired its own motor to put itself into orbit around Mars, with an eleven-hour period and a minimum distance from the surface of no more than 850 miles. Transmissions began at once, and by the time it finally lost contact, in October 1972, it had returned over 7,000 high-quality images. For the first time we could see the valleys, the gaping canyons, and the giant volcanoes, though for some weeks most of the planet was shrouded in dust, and we had to wait impatiently for the dust-storm to clear away.

The two hemispheres of Mars are not alike. Generally speaking the southern part of the planet is heavily cratered, while the northern hemisphere is lower and less cratered. There are two huge basins, Hetlas and Argyre, and there is a long, wide 'gash', Valles Marineris, which is a canyon, or rather a system of canyons, over 2,700 miles long and in places 370 miles wide and over four miles deep. But it is the volcanoes which really dominate the scene; the mightiest, Olympus Mons (Mount Olympus), three times as high as Everest, is crowned by a complex forty-mile caldera. It is a shield volcano, similar in type to our Hawaiian volcanoes, but on a much grander scale. There are three other majestic volcanoes in this part of Mars, Tharsis.

Are the volcanoes extinct? Violent eruptions belong to the past, but it has been suggested that a certain amount of activity lingers on. We will not be sure until the first astronauts arrive. At least there is no fear of dangerous 'Marsquakes', and Olypmus Mons should be easy to climb, even by a mountaineer wearing a spacesuit. The slopes are gentle, and the view from the summit will be breathtaking.

Passing over another string of Soviet failures (*Mars 4, 5, 6* and *7*) we come to the NASA *Vikings*, which were launched in August 1975. Each vehicle consisted of two parts, an orbiter and a lander. On 19 June 1976 *Viking 1* entered a closed orbit round Mars, and began to transmit pictures; the separated lander touched down gently on the 20th in the plain which had been named Chryse. The area was strewn with rocks which were obviously volcanic; the sky was yellowish-pink and, as expected, the cold was intense even in the middle of the 24.5 hour Martian day. The *Viking 2* lander touched down on the following 9 September in the more northerly plain of Utopia, with essentially similar results. At the end of the Viking missions, most people believed – perhaps rather ruefully – that Mars was completely sterile. Deliberate searches for living matter had been carried out by both landers, but with negative results.

Nothing much was done, so far as Mars was concerned, for more than a decade after the *Vikings*, and the next attempts, from 1988, were disappointing. The USSR dispatched two probes to the larger satellite, Phobos, in 1988, but both failed, one because of human error (a wrong command was sent out) and the other for 'reasons unknown'. NASA's ambitious *Mars Observer* (1992) simply 'went silent when nearing its target'. *Mars 96*, from Russia (no longer the USSR), was dumped in the sea because the third stage of the rocket launcher refused to fire. It was only with *Pathfinder* (launched December 1996) that fortunes began to change.

New methods were used. *Pathfinder* would not go into orbit round Mars, bide its time and then come down gently, using rocket braking and parachutes. It would crash through the Martian atmosphere, land, and bounce several times before coming to rest. The whole vehicle would be surrounded by airbags to cushion the shock. It sounded – it was – risky, but it worked. On 4 July 1997 *Pathfinder* landed in the Ares Vallis, bounced as expected, and suffered no damage at all. From it emerged a tiny rover, *Sojourner*, which moved slowly around, photographing rocks and analysing their materials. Contact was not lost until 7 October.

Ares Vallis is dry now, but it had once been a raging torrent, and there were rocks of all kinds. There could be no doubt that many of them had been affected by water; the Red Planet had once been warm and wet. Life could then have appeared. But had it actually done so? *Sojourner* could not say.

Several orbiters were sent in the closing years of the twentieth century. *Mars Global Surveyor* (NASA, 1998) functioned excellently for over a year. Japan's *Nozomi* (1998) never reached Mars, because of a rocket problem. NASA's *Mars Climate Orbiter* (1998) crashed to destruction because of an incredible blunder; it was programmed in imperial units, but as it prepared to enter orbit, orders were sent to it in metric! *Polar Lander* (NASA, 1999) simply 'went missing'. However, since then there have been some outstandingly successful orbiters: *Odyssey* (NASA, 2001); *Mars Express* (Europe, 2003); and *Mars Reconnissance Orbiter* (NASA, 2005). And of course there have been those two amazing rovers, *Spirit* and *Opportunity*, which have exceeded all expectations. They landed on Mars in 2004, and they were still operating efficiently four years later. Then in 2008, NASA's *Phoenix* came down in the Martian Arctic, and obtained new data about the surface layers and the underlying ice.

I have heard that someone once tried to send a telegram to Mars. Is this true?

Quite true. In 1926 Dr Mansfield Robinson went to the Central Telegraph Office in London and said that he would like to send a telegram to Mars. The postal authorities agreed, and charged him the then standard rate of eighteen pence per word, but tactfully added 'Reply Not Guaranteed'.

To the best of my knowledge, the Martians never did send an answer!

Has anyone heard radio signals from the inhabitants of Mars?

No, because there aren't any Martians – yet! But before we found that Mars does not support intelligent life, various attempts were made to 'listen out'. Two of these experimenters were world-famous scientists, Nikola Tesla and Gugliemo Marconi.

Tesla was born in 1856, in what is now Croatia. He was an eccentric and a mystic, but he was a brilliant electrical engineer whose work was of immense importance, though some of his ideas were, to put it mildly, unconventional. He emigrated to America, and while living at Colorado Springs, in 1899, he constructed his 'Magnifying Transmitter', and used it to pick up some very unusual radio signals. He wrote:

I can never forget the sensation I experienced when it dawned on me that I had observed something possibly of incalculable consequence to mankind. I felt as though I were present at the birth of a new knowledge or the revelation of a great truth. My first observations positively terrified me, as there was present in them something mysterious, not to say supernatural, and I was alone in my laboratory at night ... It was some time afterward that the thought flashed upon my mind that the disturbances I had observed might be due to intelligent control. Although I could not decipher their meaning, it was impossible for me to think of them as being entirely accidental. The feeling is constantly growing in me that I had been the first to hear the greeting of one planet to another.

The signals, he concluded, must have come from Mars. Remember that at that time many astronomers believed the Martian 'canals' to be artificial.

He gave no further details; we know little about his Magnifying Transmitter, and nobody repeated the experiment, and after Tesla's death in 1943 some of his notebooks could not be found. In 1921 no less a person than Marconi, who transmitted the first wireless message across the Atlantic, also claimed to have picked up signals from Mars, but like Tesla gave no details. In 1924, when Mars was well placed in the sky, radio stations in America stopped broadcasting for an agreed brief period so that both amateurs and professionals could 'listen out' without interference and see if they could detect any transmissions from the Red Planet. Strenuous efforts were made, but Mars remained obstinately silent!

But in recent years we have, of course, received signals aplenty from the spacecraft we have sent there. If all goes well, manned flights will take place before the end of the present century, and the first bases will be established. Then, at last, we really will receive messages from Mars.

If I go to Mars, will I be able to see shooting stars?

Yes, because although the Martian atmosphere is much thinner than ours, it is still appreciable.

Meteors dashing into the Earth's upper atmosphere become luminous when they are roughly 120 miles above ground level. The density of the Martian atmosphere decreases with altitude more slowly than ours, and at something like an altitude of eighty miles, the densities of the two are equal. Velocities are less than those of our meteors since Mars is around fifty million miles further from the Sun and the meteors will dash in more slowly, so that on average they are probably half a magnitude fainter than ours – but in the clear sky they will be striking enough.

On 7 March 2004 the *Spirit* rover, exploring the Martian surface, imaged the trail of a meteor which seems to have been from a comet known as 11 4P/Wiseman-Skiff, which has a period of 6.5 years (the name honours the discoverers of the comet, Jennifer Wiseman and Brian Skiff, who photographed it in 1986–7). From Mars the constellations look almost the same as they do from Earth, though there is a different north pole star – Deneb instead of Polaris. The meteor seen by *Spirit* apparently came from the direction of Cepheus, and there may well be Cepheids every Martian year.

There will of course be other showers, too. One, associated with the comet 79P/du Toit-Hartley, was predicted for 2003 by Dr Apostolos Christou, of Armagh Observatory, and the shower was detected, both optically and by effects in the Martian ionosphere, measured by the orbiting satellite *MGS* (*Mars Global Surveyor*).

Shooting star observers will have plenty of scope when they watch the Martian sky. Moonlight will not interfere. Phobos and Deimos will never be bright enough to flood the sky with radiance, and there is no man-made light pollution – at least, not yet! Of course, the Martian atmosphere gives no protection against meteorite larger bodies, which do not burn away during their plunge to the ground, and land intact. Meteorites have already been identified on Mars by the *Spirit* and *Opportunity* rovers. Remember, too, that Mars is not too far from the inner edge of the main asteroid zone, and there are plenty of impact craters on the planet's surface. Mars will certainly be a good vantage point from which to monitor these strange little members of the Solar System.

If we live on Mars, will our descendants be the same as us?

This is a very difficult question to answer, but differences will presumably become evident. People on Mars will live in what we would call an artificial environment, at least in the early days of colonisation, and gravity there has only one-third the strength of ours. On Earth there are striking differences between different ethnic groups, because they have evolved from a common ancestor, but under different conditions of climate; it is not usually very difficult to differentiate between an Englishman and a Frenchman. No doubt a Martian of the year AD 3000 will not look quite the same as you and me, but that is really as much as we can say at the moment. Time will tell.

I know that Mars has two moons. If you were standing on Mars, would they look as bright as our Moon does to us?

Certainly not. Our Moon is over 2,000 miles in diameter; but both the Martian moons, Phobos and Deimos, are very small. They are rather irregular in shape; Phobos has the longest diameter of less than twenty miles, Deimos less than ten. They are not original satellites, but are ex-asteroids, captured by Mars long ago. They were discovered in 1877 by the American astronomer Asaph Hall. You need a reasonably powerful telescope to see them.

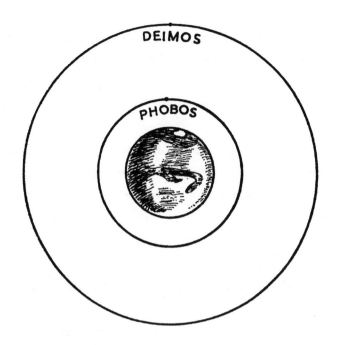

Orbits of Phobos and Deimos. Both satellites are very close to Mars

Phobos moves round Mars in a period of 7.5 hours, and is only 3,800 miles above the planet's surface. Remember, Mars spins round in just over 24.5 hours, so that anybody on Mars would see Phobos rise in the west, cross the sky in a mere 4.5 hours, and set in the east; it would do this three times a day. It would show phases, but would look much smaller than our Moon and would not be of real use as a source of light during night time. Deimos is much further out (12,000 miles) and has an orbital period of thirty hours; it would stay above the horizon for 2.5 days at a time, and would send no more light than Venus does to us. From the polar regions of Mars the satellites would not rise at all.

Phobos is spiralling very slowly down toward Mars, and will probably crash land eventually, though not for a very long time yet – at least 50,000 years. When it impacts, it will make a very large crater! Deimos is in a 'safe' orbit, and will not fall on to the Martian surface.

There have been searches for other moons of Mars, but with no success. If Mars did have an extra satellite more than a few feet across, it would certainly have been found by now.

What are Phobos and Deimos like?

We know what they are like, because they have been seen from close range by the various spacecraft that have been sent to Mars. They are cratered, because of meteoric bombardment, and there are hilts and valleys; their surfaces have loose layers, deeper on Deimos than on Phobos.

In a way these two midget moons look like space stations – and indeed a well-known Russian astronomer once suggested that they really are artificial stations, built by Martians! Alas, there are no Martians, and there is nothing artificial about Phobos or Deimos, but our future astronauts may well make use of them. I have suggested that the pioneer travellers will go first to Deimos, and establish a supply and re-fuelling base there, before making the final descent to Mars itself. (I propose Deimos rather than Phobos, because from there you would have a much better overall view.) Because their gravitational pulls are so weak, 'landing' would really be a space-docking operation, and should present no problems.

Why are the moons of Mars called Phobos and Deimos?

In Greek mythology Ares (Mars) the War God had two attendants. Phobos (Fear) and Deimos (Terror). These names for the satellites were suggested by a Mr Madan, a schoolmaster at Eton College.

If I stood on Deimos, would I be able to jump out into space?

Not quite. On Deimos (or Phobos) you would weigh very little, and if you jumped up you would take a very long time to come down, but Deimos does have a gravitational pull, admittedly a very feeble one, and you would not be able to jump clear. For a body whose density is the same as that of the rocks in the Earth's mantle, the diameter would have to be less than half a mile. With a body smaller than this, you could jump off and turn yourself into a tiny artificial satellite. (I don't recommend trying!)

You could, of course, easily throw a cricket ball or a tennis ball clear of Deimos or Phobos.

What would happen if a comet hit Mars?

It would make a crater. The nucleus of a comet is made up of ice and rocky material, so that it would be able to produce a deep hole and hurl *débris* around. No doubt this has happened often enough in the past, just as it has on Earth, and there are many impact craters on Mars, so that in the remote past the planet has been heavily bombarded not only by comets, but also by meteoroids. The American rovers *Spirit* and *Opportunity* have already sent back pictures of meteorites lying upon the Martian surface.

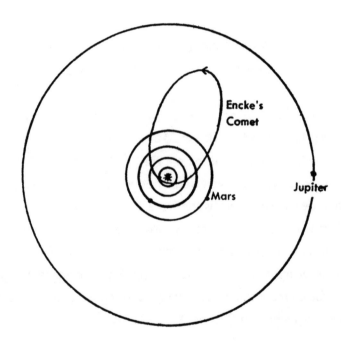

Orbit of Encke's Comet, period of 3.3 years

We sometimes see Venus and Mercury pass in transit across the face of the Sun. If I could go to Mars, would I ever be able to see a transit of the Earth?

Yes, you would; from Mars, the Earth is an inferior planet, just as Venus and Mercury are to us. If you happen to be visiting the Martian base on 10 November 2084 (our calendar!) you will see a transit of the Earth and the Moon. It will not be difficult to observe, and the Moon too will be an easy object.

From Mars, Venus will next transit on 19 August 2030; apparent diameter eighteen seconds of arc. Mercury will transit on 10 May 2013, but will look tiny, a mere six seconds of arc across. Quite possibly these events will be recorded by spacecraft near or on Mars.

Can we ever make Mars really like Earth?

This process is called 'terraforming', and not everyone believes that we should ever attempt it. It will be immensely difficult, and will take centuries, if indeed it can be managed at all.

On Mars, the first thing to do must be to thicken its atmosphere, and make it fit to breathe. If this can be done, the surface temperature would automatically rise. We must however remember that any new atmosphere will leak gradually away, just as the original atmosphere did, so that it will have to be continually replenished; 'greenhouse' gases, so unpopular on

Earth, will be very welcome on Mars. Every effort will be made to develop plants which will grow on Mars, and will help in changing the atmosphere as well as providing food.

A supply of water is essential, but there should be no problem here, because there is plenty of H_2O in the form of ordinary ice. Much more serious is the fact that the present Martian atmosphere is unlikely to give full protection against harmful radiations from the Sun in space, and quite apart from the inconvenience to colonists, who would have to retire to radiation shelters at times of high solar activity. Future plants will need to be shielded. All in all, nearly everything depends upon modifying the atmosphere.

We are not yet able to consider terraforming; before we think about terraforming Mars, we must learn how to get there! But if we make our first trips before the mid century, and set up bases before 2100, we might start the process before 2200. By AD 3000 Mars might again be green and welcoming, with water flowing in canals. I wonder if it will really happen?

Can you play cricket on Mars?

There is no reason why not, but there are obvious problems.

First, if you are in the open you will have to wear full space clothing, which will be very clumsy. Secondly, you will be under only one-third Earth gravity, and running would be awkward. You have probably seen TV pictures of the *Apollo* astronauts on the surface of the Moon, and noted that everything seems to happen 'in slow motion'. The same will be true on Mars, though to a lesser extent. I am a leg-spinner but with an unusually long run (fourteen paces) and I do not know how I could manage; a fast bowler would be even worse off; the thin Martian atmosphere could not make the ball swing through the air. On the other hand, the batsman could hit the ball a very long way, and a cricket ground would have to be vast! A prepared wicket would be essential, because the natural surface of Mars would be hopelessly unsuitable.

Inside a Martian base, with Earth-type atmosphere and temperature, things would be easier, but a lofted drive would hit the roof; there would have to be a special rule about this. Also, the base would have to be very spacious.

No doubt cricket will come to Mars one day, and it will be interesting to see whether the colonists on the Red Planet will be able to raise a team good enough to beat the Australians. And will they ever produce a player to match W.G., Don Bradman or Shane Warne? Time will tell.

Why did astronomers originally think that there ought to be a planet whose orbit lay between those of Mars and Jupiter?

It is very obvious that the Solar System is divided into two main parts. We have the rocky inner planets, from Mercury to Mars, and then a wide gap before coming to the giants; on average Mars is just over 141 million miles from the Sun, Jupiter as much as 483 million miles. An eighteenth-century German astronomer named Titius found what he thought was a strange mathematical relationship linking the distances of the planets; it was popularised by a much better-known astronomer, Johann Elert Bode, and is always known as Bode's Law.

Astronomers took it very seriously. It indicated that there should be a planet in the Mars-Jupiter gap, and in 1800 a number of observers decided to search for it. They called themselves the 'Celestial Police', and began work under the direction of Johann Schröter and the Baron von Zach. Actually, the first discovery was made by an Italian astronomer, Piazzi, who was not then a member of the 'Police' (though he joined later). On 1 January 1801, the first day of the new century, he discovered the expected planet. It was small – obviously much smaller than the Moon – and was much too faint to be seen with the naked eye, but at least it was in the right position. It was named Ceres, after the Earth goddess.

The 'Police' were not satisfied, and continued the hunt. By 1808 they had picked up three more small worlds: Pallas, Juno and Vesta. The four became known collectively as the Asteroids. The 'Police' gave up the search in 1815, and no more were found until 1845. Today tens of thousands are known, though there are none as large as Ceres (diameter 603 miles) and only Vesta is visible with the naked eye.

Is Bode's Law really valid?

I think the answer must be 'no'. It works fairly well for the planets out to Uranus, but it breaks down completely for Neptune, the third most massive planet in the Solar System. There seems no reason for it, and almost all astronomers now dismiss it as having no real significance.

Has anybody ever been to an asteroid?

No, but on 12 February 2004 an unmanned spacecraft made a controlled landing on asteroid 433 Eros, which is shaped like a lozenge, with a longest diameter of 20.5 miles and a width of eight miles. Eros was discovered in 1898, and was the first known asteroid to cross the orbit of Mars. It can occasionally come within fifteen million miles of the Earth, though there is no chance that it will hit us.

In February 1996 a spacecraft was sent to it; the vehicle was originally called *NEAR* (*Near Earth Asteroid Rendezvous*) but was then renamed in honour of the American astronomer Eugene Shoemaker. It was meant to survey the asteroid from orbit, but it was found possible to touch down gently; it did so, and continued to transmit for over two weeks. Eros is solid and rocky, with craters, valleys, hills and ridges. The rotation period is five hours, and the temperature fluctuates wildly; 100 degrees celcius in the 'daytime', plunging to minus 150 degrees celcius at night, so that spending a few hours there would certainly be an unusual experience.

NEAR-Shoemaker is still sitting where it landed, in a shallow depression which has been named Himeros. There it will stay until a future astronaut collects it – but I feel that it would be better left *in situ*, as a sort of Solar System historical exhibit!

Has a spacecraft ever landed on an asteroid and then taken off again?

Yes, once. The asteroid was 25143 Itokawa, and the spacecraft was Japan's *Hayabusa*.

Hayabusa was launched on 9 May 2003, and reached its target in November 2005; Itokawa is very irregular in shape, with a longest span of about three miles. The original plan was to fire tiny projections at the asteroid, collect the resulting 'spray' in a collection 'horn' and then return home with the samples. Unfortunately things went wrong, and two attempts with the projectiles failed. Nothing daunted, the Japanese controllers managed to bring *Hayabusa* down gently on the surface (19 November), where it remained for half an hour before taking off again for the return journey. It is hoped that some of Itokawa's surface material was picked up in the sampling chamber.

The problems were far from over, and as I write these words (June 2008), the only other asteroid to be reached so far is 433 Eros, by the *NEAR-Shoemaker* probe (see p. 99); but this was a one-way journey only, and was never intended to be anything else.

Is there a coal-black planet in the Solar System?

Not coal-black, but asteroid 253 Mathilde is certainly very dark. It reflects only about three per cent of the sunlight falling on it so that its albedo (reflecting power) is about the same as that of fresh asphalt. It has been said to be 'blacker than coal'. It is unusual in other ways, too.

Mathilde was discovered on 12 November 1885 by the Austrian astronomer Johann Palisa, Director of the Vienna Observatory (it was named after Mathilde, wife of Moritz Loewy, Director of the Paris Observatory). It is irregular in shape, with a longest diameter of thirty-three miles; it moves round the Sun in the Main Belt of asteroids, so that it stays in the region between the orbits of Mars and Jupiter. Its orbital period is 4.3 years (1,573 Earth days) but its own 'day' is very long – 17 days 9 hours 45 minutes (415 hours). Of all known asteroids, only 288 Glauke and 1220 Crocus spin more slowly than that.

The whole surface is heavily cratered, so that the asteroid must have been mercilessly bombarded. We know a great deal about it, because on 27 June 1997 the *NEAR-Shoemaker* spacecraft, on its way to rendezvous with asteroid 433 Eros, flew past it at a range of only 748 miles, and sent back over 500 images in less than half an hour. Because Mathilde is so dark, it seemed fitting to name the craters and basins. Here are a few of them:

Karoo (diameter 20.7 miles), South African coal basin.

Kuznetsk (16.1), Russian coal basin.

Damodar (17.8), Indian coal basin.

Jixi (11.7), Chinese coal basin.

Matanuska (1.6), Alaskan coalfield.

Oaxaca (3.2), Mexican coal basin.

Clackmannan (1.7), Scottish coal basin.

Otago (4.9), New Zealand coalfield.

But don't expect to find coal there – so you can't go mining. Moreover, Mathilde is a very cold place!

Do asteroids ever collide with each other? If so, would it be dangerous to go into the asteroid zone?

Yes, there are collisions even now, and many of the very small asteroids are fragments which have been broken away from larger ones. Collisions must have been even more in the past. But if you take a spacecraft into the main zone, it is very unlikely that you would be hit. After all, quite a number of spacecraft have been right through the whole swarm without being damaged. Obviously we can't be sure, and accidents will happen, but the danger in the main belt of asteroids does not seem to be as was once feared.

Is there a belt of asteroids closer to the Sun than the orbit of Mercury?

There could be. We call these asteroids 'Vulcanoids', after the planet that was once thought to exist there, and was even named Vulcan, after the blacksmith of the gods. There is no large planet there – no Vulcan – but a belt of asteroid-sized bodies could well exist.

The trouble is that they will be very difficult to locate, even if they really exist. Their gravitational pulls will be too weak to affect any bodies which we can see; they will be too small to be detected passing in transit across the solar disk, and they will always be so close to the Sun that they will be drowned in the glare. Energetic searches are being made, and though there has been no success as yet I hope it will not be long before we track down our first Vulcanoid.

If I landed on an asteroid, could I walk around?

It wouldn't be very easy. Even on Ceres, which is by far the most massive member of the swarm, you would weigh very little, because its gravitational pull is so weak – and on a very small asteroid you could jump clear altogether, turning yourself into a tiny independent satellite of the Sun! We know of asteroids with diameters smaller than the length of a cricket pitch (or a baseball field).

I know that the planets are named after mythological gods – Mars, Venus, Jupiter, Saturn, Mercury, Neptune – but what about Uranus?

In mythology, Uranus – Ouranos, properly pronounced 'Oo-ran-os', not 'Yew-ray-nus' – was Saturn's father and was the first ruler of Olympus. He was overthrown by Saturn, who was in turn overthrown by his son, Jupiter. The Olympians do not seem to have been a close and affectionate family!

The original names were of course Greek, but we always use the Latin versions, so that, for instance, Zeus has become Jupiter, while Ares is known as Mars.

Is Jupiter's red spot a volcano in eruption?

Not if it is a whirling storm – a phenomenon of Jovian 'weather'. It has been seen on and off (more on than off) ever since the sixteenth century, and it has been known to be 30,000 miles long, as it was in 1878. Its longevity is due to its exceptional size.

A second red spot, seen since 2007, is about half the size of the first. The cause of the red colour is unclear, but it may be due to phosphorus dredged up from deep inside Jupiter.

Was Saturn ever thought to
be a miniature sun?

Indeed it was. Both Jupiter and Saturn were believed to send out enough heat to warm their satellite systems, and no doubt this was also true for Uranus and Neptune. Some remarkable conclusions followed, as described by the well-known astronomer R.A. Proctor in his book *Our Place Among Infinities*, published in 1875. He knew that Saturn had eight satellites. Titan was by far the largest, and then Iapetus (he used the alternative spelling, Japetus) which he believed to be much bigger than it really is; Titan was thought to be comparable in size with Mars. Like many others – even astronomers – Proctor was inclined to believe that satellites had been created by the Almighty specifically for the benefit of the inhabitants of their planets, but he knew that Saturn did not have a visible solid surface, and in any case was likely to be unpleasantly hot. Moreover, the satellites would be dimly lit, because Saturn is so far from the Sun, that they could not provide much light. So why were they there at all? This is what Proctor had to say about it:

> Certainly, if we consider what the Saturnian satellite family really
> is, that the orbs composing it are all large in reality, however
> minute they may appear either when viewed with the telescope
> or when considered with reference to such orbs as Jupiter or
> Saturn; that the span of the complete system is no less than
> 4,400,000 miles, or more than five times the Sun's diameter;
> that even Japetus, which moves the slowest, circles on his orbits

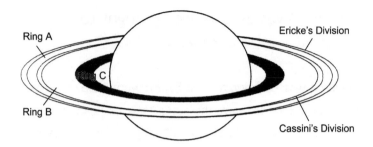

The ring system of Saturn

with a rapidity which exceeds a hundred-fold the swiftest of our express trains – we cannot but regard this system of secondary orbs as a most important portion of the scheme ruled over by the Sun. If we are compelled to believe that the purpose intended to be fulfilled by these bodies is not the illumination of the Saturnian nights – and for my own part I can arrive at no other conclusion – we seem bound to believe that they were created for some other purpose of importance. It does not seem at all unlikely, on this view of the subject, that they are themselves the abode of living creatures of various orders. I have before shewn reasons for believing that Saturn may be a source whence heat is supplied to these eight orbs, whereas it seems unlikely that he is himself a world fit to be the abode of living creatures. Again, though the satellites supply Saturn with very little light, yet they are capable of supplying each other with no inconsiderable amount, and must frequently present phenomena of great beauty and interest as viewed from each other. Thus a variety of reasons suggest the probability that we are to look among the Saturnian satellites, and not to Saturn himself, for places fit to be the abodes of living creatures.

Well – we know about Titan, with its methane lakes; Enceladus, with its ice fountains; Iapetus, with its towering equatorial ridge – but much though I would like to meet a Iapetan or an Enceladan, I do not regard it as very likely. Pity!

Titan has seas on its surface. Could fish live in them?

I very much doubt it! The seas and lakes of Titan are not in the least like ours.

Titan is the largest satellite of Saturn, and it really is big; its diameter, over 3,000 miles, is slightly greater than that of the planet Mercury. It is the only planetary satellite to have a substantial atmosphere, and in fact Titan's atmosphere is much denser than that of the Earth. It is made up largely of nitrogen, together with a good deal of methane, which is a poisonous hydrogen compound (it is often called 'marsh gas'). The thick atmosphere hides the surface completely, so that before the space probes were developed we knew nothing about what lay underneath. We were not even sure whether or not the surface was covered with water or some other liquid.

The first real information came from the *Cassini* spacecraft, which surveyed the whole of the surface and also dropped a small lander, *Huygens* (named after the Dutch astronomer Christiaan Huygens, who discovered Titan in 1656). *Huygens* came down successfully, and transmitted pictures and data for some hours after arrival – an amazing feat when you remember

Liquid Lake on Saturn's Titan. Photographed by *Cassini* Radar Mapper, JPL, ESA, NASA

that Saturn is almost 900 million miles from the Sun. Predictably, Titan proved to be bitterly cold – far too cold for surface water to exist. *Huygens* came down on solid but 'squashy' ground, but there were obvious signs of river channels. Liquid flowed there, but it was liquid methane, and methane rain was common. In fact, methane is Titan's equivalent of water.

After *Huygens* fell silent, the *Cassini* spacecraft continued orbiting Saturn, and sent back clear images of Titan showing a system of lakes and channels – a veritable Lake District. But the lakes were found to be filled with a mixture of liquid methane and another hydrogen compound, ethane. Bearing this in mind, as well as the intense cold, we can be sure that no fishes of the kind we can understand will be found there.

This does not necessarily rule out all living things; after all, life is very adaptable, and we find it in places as unwelcoming as our own undersea hydrothermal vents. To claim that Titan is absolutely sterile would be far too sweeping, and this strange world may have many surprises in store for us. However, it is hard to picture future astronauts planning a pleasant afternoon's fishing in the Lake District of Titan!

Is it true that one King of England was deeply interested in astronomy?

Quite true. He was George III, who was also King of Hanover.

In 1769 there was a rare event: a transit of Venus, when the planet passed across the face of the Sun. The king was anxious to see it, and a small observatory, equipped with a telescope, was built at Kew, in outer London. It is still there, though the building is no longer used as an astronomical observatory.

In 1781 the planet Uranus was discovered by William Herschel, a young Hanoverian musician who had settled in England and had become organist at the Octagon Chapel in the fashionable resort of Bath. The king was intrigued; he gave Herschel the title of King's Astronomer, arranged for him to move to Slough, near the royal residence at Windsor, and granted him a pension so that he could devote all his time to astronomy. Herschel subsequently became one of the greatest of all astronomical observers.

Why did astronomers once think that there were no planets beyond Saturn?

For rather a curious reason. Five planets – Mercury, Venus, Mars, Jupiter and Saturn – had been known since very early times, because they are naked-eye objects, and move around against the background of stars. Add the Sun and Moon, and you have a total of seven. Seven was the mystic number of the ancients, so clearly there must be seven members of what we now call the Solar System. This seemed eminently satisfactory, and nobody really thought about another planet until 1781, when William Herschel discovered Uranus.

Herschel was not looking for a planet. He was Hanoverian, but came to England as a young man, and spent the rest of his life here. He was an expert musician, and became organist at the Octagon Chapel in the fashionable resort of Bath, where he was a great success both as a performer and a teacher; with him was his devoted sister Caroline. He became interested in astronomy, made his own reflecting telescopes (which were the best of their time) and started to 'review the heavens', mainly because he wanted to find out how the stars were arranged in space. On 13 March 1781, using one of his telescopes from the garden of his house in New King Street, Bath, he found an object which he knew could not be a star, because it showed a small disk, and it moved slowly from night to night. He thought that it must be a comet, but when the orbit was worked out it proved to be a planet, moving round the Sun far further out than Saturn. It was named Uranus, in mythology, after Saturn's father, the first ruler of Olympus. It is only just visible with the naked eye, which is why nobody had identified it before Herschel did so; several astronomers had recorded it, but had always mistaken it for an ordinary star.

How long is a year on Uranus?

Uranus takes eighty-four Earth years to complete one journey round the Sun, so that Herschel discovered it less than three Uranian years ago. But the axial rotation period is short; only 17 hours 14.4 minutes. This means that there are almost 43,000 days in every Uranian 'year'.

The planet's calendar is made even more complicated by the fact that the rotational axis is tilted to the orbital plane by ninety-eight degrees – more than a right angle. Our tilt is 23.5 degrees, which is why we have our seasons, but Uranus 'rolls' along; sometimes one or other of the poles is pointed directly toward the Sun, and a polar night can last for twenty-one Earth years. We do not know why Uranus is tilted in this extraordinary way. It used to be thought that long ago Uranus was hit by a massive impactor and literally knocked sideways, but this is rather difficult to believe, and more probably the tilt was gradually increased by gravitational interactions with the other planets. But why should Uranus be so unlike the other giant planets in this respect? Again we have to confess that we do not know. In some ways Uranus is very much of an enigma.

Could I go for a walk on Uranus?

No, because the surface is not solid; it is made up of gas. Deep down there could be a silicate core, but we cannot be certain; we think that most of the globe is made up of 'ices', including water ice, while the outer gas – Uranus' atmosphere, if you like – consists mainly of hydrogen, together with some methane and ammonia. Methane absorbs the red part of the sunlight falling on the planet, so that Uranus looks pale green.

It is a giant, 30,000 miles across (rather less than half the diameter of Saturn), but is much less dense than our world. Put Uranus in one pan of a huge pair of scales, and you would need only fourteen Earths to balance it. It is worth noting that Uranus, unlike the other giant planets (Jupiter, Saturn and Neptune) does not seem to have an internal heat source; just why this is so, we know not.

I know that Jupiter is the largest planet in the Solar System, but which is the smallest? Is it Pluto?

Pluto is indeed small – smaller than our Moon, but Pluto is no longer classed as a planet; it is a dwarf planet. This means that the smallest planet is Mercury, whose diameter is 3,033 miles. Actually there are two planetary satellites, Ganymede in Jupiter's system and Titan in Saturn's, which are slightly larger than Mercury, but they are less dense and less massive.

We are now sure that we know all the planets in the Solar System, and of these Mercury is quite definitely the smallest.

Could I go for a walk on a comet?

No, I'm afraid not. As I've said, the only really solid part of a comet is the nucleus, which is never more than a few miles in diameter, and it is not very dense — it's made up of rocky fragments together with ice, so that the mass is very low. This means that the pull of gravity is very feeble; it would be possible to land on a comet, but you would be virtually 'weightless', so that ordinary walking would be out of the question. If you stood on the surface and jumped up, you wouldn't come down; you would leap clear of the comet, and become an independent spacecraft!

When will I next see Halley's Comet with the naked eye?

In 2061. The comet has a period of seventy-six years, and was last at perihelion in 1986. It is the only periodical comet which can ever become bright.

A conspicuous comet was observed in 1682 by Edmond Halley, later the second Astronomer Royal. He calculated the orbit, and realised that it moved in the same path as comets previously seen in 1607 and in 1531; if these three were identical, the next return should take place in 1758. On Christmas Night

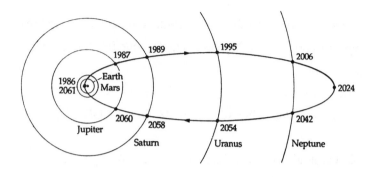

Path of Halley's Comet

of that year the comet was found by a German astronomer, J. Palitzsch, and it returned to perihelion in 1759. Fittingly, it was named in honour of Edmond Halley.

Records of it were then traced back, dating to Chinese times. In AD 837 it was at its very best, with a head as bright as Venus and a tail extending over ninety-three degrees of the sky; it was prominent in 1066, before the Battle of Hastings – it is shown on the famous Bayeux Tapestry, with King Harold tottering on his throne. In 1301 it was seen by the painter Giotto di Bondone, who used it as a model of the Star of Bethlehem in his painting 'The Adoration of the Magi' – though we may be sure that what the Star of Bethlehem was, if anything, not Halley's Comet. At the return of 1456 it was again regarded as an evil omen, and the Pope, Calixtus III, preached against it as an agent of the Devil, though the comet appeared to be quite unfazed.

Since Halley's time the comet has returned in 1835, 1910 and 1986. In 1986 it was badly placed, and was not a bright naked-eye object, but several spacecraft were sent to it, and one of these, Europe's *Giotto* (named after the painter) went right into the head, and obtained images of the nucleus from reasonably close range. At present it is on its way back to aphelion, well beyond the orbit of Neptune, but it will be back at perihelion in 2061. I am afraid I will not be there to welcome it, unless of course I live to the advanced age of a hundred and thirty-eight, but if you observe it please give it my best wishes.

It will not be brilliant in 2061, but in 2137 it should be truly magnificent. By then there will no doubt be expeditions to it in both manned and unmanned spacecraft.

Could I buy a bottle of Comet Wine?

I doubt it, but I do know what you mean.

Some years ago I had a letter from Sotheby's, the auctioneers. They had a bottle of 'Comet Wine' for sale – did I know anything about it? I did, and I told them.

In 1811 a 'great' comet appeared; it was discovered by the French astronomer Honoré Flaugergues on 25 March, and became visible with the naked eye in broad daylight. The comet was 120,000 miles in diameter, and the tail extended over ten million miles. In that year the wine crop in Portugal was particularly good, and the growers attributed this to emanations from the comet! For years afterwards Comet Wine appeared in the price lists of wine merchants.

When I had this letter from Sotheby's in 1984, I asked them to take a photograph of the bottle and send it to me. They did.

I imagine the bottle was then sold, but I hate to think what the wine would have tasted like if anyone had been brave enough to try it!

Can a comet simply vanish without trace, and for no apparent reason?

Yes, it can – and it has happened recently, much to the annoyance of NASA.

The story begins with the *Deep Impact* space mission, in 2005. This was a 1,450lb probe, launched on 12 January to rendezvous with well-known periodical comet, Tempel 1. It would fly past the comet, and drop a 770lb lander, made chiefly of copper, which would hit the surface and produce a crater at least thirty feet across – it would not do any serious damage to the comet, but would allow NASA scientists to examine what lay just below the surface. Minutes later the fly-by section of the spacecraft would pass by at a range of 300 miles, taking photographs to show what had happened. The entire event would also be imaged with telescopes on Earth as well as the orbiting probe out as far as Uranus. It was not a bright comet, but it suited NASA very well, and preparations for the encounter began well ahead of time. Telescopes were from observatories at Chandra, Spitzer, Hubble and XMM-Newton.

All went well. The impactor came down exactly as planned, on 29 June; a crater was formed, and the results were of immense interest – for example, the *débris* ejected contained more dust and less ice than had been expected. The comet was by no means substantial – it was about seventy-five per cent empty space, and there was very fine material more like talcum powder than sand. Quite unfazed, Tempel I continued its placid journey round the Sun. The fly-by section of Deep Impact had other work ahead; it was scheduled to go on to survey another comet, 85 P/Boethin, on 5 December 2008. It would not release another lander, but would take photographs from close range.

Boethin's Comet had been discovered in 1975 by a clergyman, the Rev. Leo Boethin, and found to have a period of 11.3 years; at perihelion it was 101 million miles from the Sun – just outside the orbit of the Earth, while at aphelion it swung out as far as Uranus. Telescopes made ready to photograph the comet. But – where was it?

It simply wasn't where it ought to have been. There could be no conceivable error in the calculated position; the trouble lay with the comet. Eventually NASA had to admit defeat, and re-directed Deep Impact to another comet, Hartley 2.

What had happened to Boethin? Had something hit it and broken it up? Suggestions that some malevolent demon had snatched it out of the sky were not well received, so the only reasonable answer was that it had gently dissipated. At any rate, it has never been found. Hartley 2 is further away, and cannot be encountered before 2010, but there was really no alternative.

Well – what can one say except 'Please may we have our comet back?'

When will I next be able to see Comet Arend-Roland, which was bright in April 1957?

Sadly, the answer is 'never'. I am particularly sorry about this, because it was the subject of my very first *Sky at Night* programme on BBC television.

It was discovered by two Belgian astronomers, and became quite conspicuous. It certainly cannot be ranked as a 'great' comet, but its nucleus reached the first magnitude, and there was a long tail. In fact it looked as if there were two tails, one pointing away from the other and the other directly towards it. Actually the sunward spike was not a tail at all, but merely *débris* left behind by the comet as it travelled through space, but the effect was both interesting and unusual.

Comet Hale-Bopp which I photographed when it was bright. Note my astronomical weather vane!

Arend-Roland passed through perihelion, and then faded as it began its outward journey from the Sun. Normally it would have returned to the Oort Cloud, from whence it had come, swung round, and started on its long journey back to its next aphelion. But it was not to be. It passed by Jupiter and the gravitational pull of the Giant Planet threw it into a hyperbolic orbit. A hyperbola is an open curve, so that Arend-Rotand will never come back. It has been expelled from the Solar System.

I wonder just where it is now? Far away, wandering through 'deep space'. I remember it with deep affection, and I wish it well!

Have comets ever been seen to break up?

Yes, indeed they have. The classic case was that of Biela's Comet which had a period of 6.5 years. It was discovered in 1772, and came back regularly; at the return of 1826 it was first seen by the Austrian amateur von Biela after whom it was named. (It was independently discovered some nights later by a French observer, Gambart, and the French always called it Gambart's Comet, but nobody else did.) At the return of 1846 it split in half. The pair came back on schedule in 1852, but have never been seen since, and have certainly disintegrated, though subsequently meteors were seen coming from the position where the comet ought to have been. These meteors seem to be absent now, in which case we really have said goodbye to Biela's Comet.

Much more recently another periodical comet, Schwassmann-Wachmann 2, has been showing obvious signs of breaking up, and cannot be expected to last for much longer. And Boethin's Comet, which was to be the target of a NASA space probe in 2003, has simply 'gone missing'. What has happened to it we do not know, but the NASA probe was hastily re-directed to another comet. There was no point in chasing a ghost!

Every August we see shooting stars – the Perseid shower. If I go out to watch them, am I in any danger of being hit by one?

No. Shooting star meteors are cometary *débris* (the August Perseids are from Comet Swift-Tuttle). They are the size of a grain of sand and burn away when still forty miles above the ground, finishing their journey in the form of fine dust. Meteorites, much larger bodies, usually irons or stones, land intact and may produce craters, but are not associated either with comets or with meteors; they come from the asteroid belt.

What is the 'false dawn'?

This is an old name given to the phenomenon we now call the Zodiacal Light. When seen in the morning sky it may seem to indicate the approach of sunrise when, in fact, the Sun is still some way below the horizon.

There is a vast amount of fine 'dust' in the Solar System, spread mainly along the ecliptic, which passes through the constellations of the Zodiac. When the Sun shines on this dust between the Sun and the Earth the dust is illuminated by forward scatter, and we see the glow, either for the period before sunrise or after sunset. You need a dark, clear sky, so that the light cannot be seen for long; the Sun itself does need to be well out of view. It is best seen (in the evening) in northern spring or in autumn (in the morning), because the ecliptic is then at its closest to being perpendicular in the sky. The Zodiacal Light was described and correctly explained by the Italian astronomer G.D. Cassini in 1683, but the Persians and the Arabs had seen it much earlier than that.

What about the dust lying on the far side of the Earth with respect to the Sun? This also is illuminated, but we see it by back scattering rather than forward scattering. The result is the 'Gegenschein' (Counterglow), a faint patch of light exactly opposite the Sun. It is always very elusive; look for it when the Sun is well away from the Milky Way, i.e. from February to April and September to November, when the anti-Sun position is high at local midnight. From England I have seen it only once, in March 1942, when the whole country was blacked out as a precaution against German air raids. I was briefly on leave from the RAF, and the sky was velvety black, as I saw it when I was flying.

There is also the Zodiacal Band, a dim, parallel-sided band of radiance which may extend to either side of the Gegenschein, or be prolonged from the apex of the Zodiacal Light to link up with the Gegenschein. I have never seen it really well.

Does the Sun have any other name?

Of course every language has its own name for the Sun, but the Latin was 'Sol' – hence our adjective 'solar'.

Is it true that people were once forbidden to believe that the Sun was the centre of the Solar System?

Quite true. The Earth was believed to lie in the exact centre of the universe, with everything else – including the Sun – moving round it, completing one full circuit every day. This was strongly supported by the Christian Church, and daring to suggest that the Earth could be an ordinary planet, moving round an ordinary star, was condemned as sheer heresy. It was also said that all bodies must move round the Earth in orbits (paths) which were circular, because the circle is the 'perfect' form, and nothing except absolute perfection can be allowed in the heavens.

However, this did not fit the observational data. The Greek astronomers of over two thousand years ago knew that the planets do not move as they would do if they travelled round the Earth in circular orbits, and they worked a system according to which each planet moved in a small circle or 'epicycle', the centre of which (the 'deferent') itself moved round the Earth in a perfect circle. This picture was worked out in detail by Ptolemy, last of the great Greek astronomers (AD120–180) and is always known as the Ptolemaic theory, though Ptolemy himself did not actually invent it.

It could be made to fit the observations, but it was hopelessly clumsy and artificial. Over a thousand years after Ptolemy's time a Polish mathematician, Copernicus, realised that most of the difficulties could be solved simply by removing the Earth from its proud position in the centre of the planetary system and putting the Sun there instead. His theory was published in 1543, and the Church was bitterly hostile. Copernicus – himself a canon – had expected this, and had withheld publication of the theory until he was dying. Some of his followers were not so prudent, and one of them, Giordano Bruno, was condemned by the Inquisition in 1600 and was burned at the stake in Rome. His open support for the Copernican theory was not his only crime in the eyes of the Church, but it was certainly a serious one.

Support for Copernicus grew steadily, despite the Vatican. In the early part of the seventeenth century telescopes were invented, and in 1610 the Italian scientist Galileo made observations showing that the Ptolemaic theory could not possibly be correct. He too was brought to trial in Rome, and forced to publicly 'curse, abjure and detest' the heretical idea that the Earth moves round the Sun, after which he was kept under house arrest for the rest of his life (he died in 1642). But in England, safe from the power of the Catholic Church, Isaac Newton worked out his Laws of Gravitation, upon which all later work has been based. Newton's great book, known to us as the *Principia*, was published in 1687, and settled the argument once and for all. After Newton, the Ptolemaic theory was abandoned by all serious scientists, and the Earth was relegated to its true status as a minor member of the Solar System.

In 1992 the Pope finally admitted that Galileo had been right all along, and officially pardoned him. It had taken the Church well over three centuries to make up its mind!

Do we have summer when the Earth is at its closest to the Sun?

Not in the northern hemisphere of the Earth. Our orbit round the Sun is not circular; it is elliptical. The Sun occupies one focus of the ellipse, while the other focus is empty. We are 94.5 million miles from the Sun at aphelion, around 4 July, and only 91.5 million miles away at perihelion, around 2 January. (These are the 2008 dates; they vary slightly, because of the rather uneven nature of our calendar; for instance, leap years have to be taken into account.) The seasons are due to quite a different reason, and our changing distance from the Sun has surprisingly little effect on them. After all, the Earth's path is not so very different from a circle; a range of three million miles is not much when the mean distance from the Sun is ninety-three million miles.

The Earth's axis of rotation is not perpendicular to the plane of the orbit; it is inclined at an angle of 23.5 degrees. For six months in the year the Earth's northern hemisphere is tilted toward the Sun, and this is when we in Britain enjoy our summer; it is winter in Australia. For the other six months conditions are reversed.

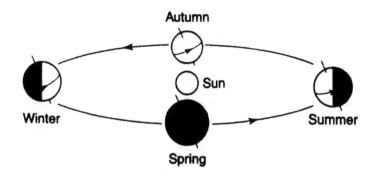

Autumn

Sun

Winter

Summer

Spring

The seasons in the northern hemisphere

The Earth completes one journey round the Sun in 365 days (more accurately, 365.25 days) and this is also the time taken for the Sun to go right round the sky against the starry background. Of course we cannot see the Sun and the stars at the same time, because the sky is so bright, but we can work out the Sun's position at any time. In a year it passes through the constellations of the Zodiac, it crosses the celestial equator twice, once around 21 March, moving from south to north (Vernal Equinox) and the other around 22 September, moving from north to south (Autumnal Equinox). Again the calendar dates vary slightly; in 2008 the equinoxes fell on 20 March and 22 September. At the time of the Vernal Equinox the Sun lay in the constellation of Aries (the Ram) when these names were given, over 2,000 years ago, and we still talk about the First Point of Aries, but slight changes in the inclination of the Earth's axis has now shifted it into the adjacent constellation of Pisces (the Fish), while the Autumnal Equinox has shifted from Libra (the Scales) into Virgo (the Virgin).

I know that light travels very quickly. How long does it take to travel from the Sun to the Earth?

Light moves at 186,000 miles per second, and on average the Sun is ninety-three million miles from us (of course these numbers are rounded off; we know the actual values very precisely). It therefore takes sunlight 8.3 minutes to reach the Earth. Look at the Sun, and you see it not as it is 'now', but as it used to be 8.3 minutes ago. If some malevolent demon suddenly snatched the Sun out of the sky, it would be 8.3 minutes before we realised that anything unusual had happened. (Fortunately, it seems that the present age is the close season for wandering demons.)

Why is it so dangerous to look at the Sun through a telescope?

Because you will certainly damage your eye, and may well blind yourself permanently.

The Sun sends out radiations at all wavelengths. We see only those between the red and violet parts of the electromagnetic spectrum, but all the rest are there, and a telescope will bring some of them to focus on your eye, with disastrous results. In any case, you can imagine the 'dazzle' effect!

This however is only a small part of the problem. The Sun is very hot, and to focus the heat on to your eye will destroy your optic nerve at once. Even a glance lasting for a fraction of a second will be very damaging. There is an easy way to demonstrate this. Next time you have telescope available, and the Sun is shining, point the telescope at the Sun, **without putting your eye anywhere near**, and then hold a piece of paper behind the eyepiece. It will at once start to smoke, and will then burst into flame. Exactly the same thing would happen to your eye.

Even looking straight at the Sun with the naked eye is emphatically not to be recommended, and remember that ordinary dark glasses do not help, because they still transmit the harmful waves. Unfortunately, some small telescopes are sold with dark 'sun-caps', the idea being to screw the cap over the eyepiece for direct viewing – **never use these**. They give no real protection, and the heat makes them liable to splinter without warning, so that the viewer will not have enough time to get his eye out of the way.

All this being so, how can we use a telescope for solar work?

There are various filters and 'wedges' which are safe, but these must always be obtained from specialist suppliers, and must never be used unless you know exactly what you are doing. But the only really sensible method is that of projection. Aim the telescope at the Sun, again keeping your eye well away from the danger zone, and hold or fix a screen a few inches behind the eyepiece (a sheet of cardboard will serve quite well). With a little adjustment, the image of the Sun will be clear, with any sunspots that happen to be around.

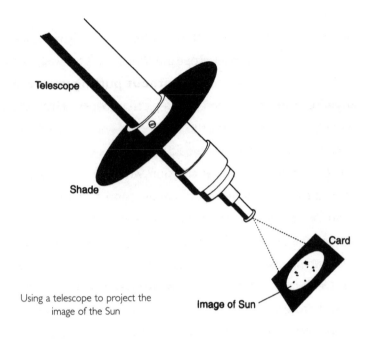

Telescope

Shade

Card

Image of Sun

Using a telescope to project the
image of the Sun

Always take the greatest care: the Sun is **dangerous**, and things can easily be overlooked. (I remember, for example, the case of a friend of mine who was projecting the Sun with his three-inch refractor, and forgot to cap his finder; he suddenly realised that his beard was alight!) In short, there is one golden rule about looking straight at the Sun with a telescope, or binoculars: **don't**.

When the Sun is low down, is it safe to look straight at it through binoculars?

No. Even when the Sun is reddened by horizon mist, the dangerous rays can still pass through. Don't risk it.

Is the Sun burning? If not, how does it shine?

The Sun is not burning in the usual sense of the term. A sun made up entirely of coal, and burning as fiercely as the real Sun, would turn to ashes in a mere 4,600 years, and we can show that this theory must be wrong, because it is known that the Earth has existed for 4.6 thousand million years, and the Sun is certainly older than that.

Various theories were proposed. According to one of these, the Sun was kept hot by being constantly bombarded by rocks from space. A better was due to a German astronomer, von Helmhottz, in 1854; he believed the Sun to be shrinking, so that its diameter decreased by about 200 feet per year. The material below the surface would be compressed, and would produce enough heat to keep the Sun shining; this would suffice for around fifteen million years. But the true answer was found by Hans Bethe in 1938 (others were on the same track, notably George Gamow and Carl von Weizsäcker): the Sun produces its energy by using hydrogen as fuel.

As we have noted, the Sun is made up largely of hydrogen. Deep down, where the temperature is so high and the pressure is colossal, the hydrogen atoms (or rather the hydrogen nuclei) are combining to form nuclei of a heavier element, helium. Four hydrogen nuclei are needed to make one helium nucleus, but in the process a little mass is lost and a little energy is released. The loss in mass (or 'weight', if you like) amounts to four million tons per second; the Sun weighs much less now than it did when you picked up this book. The energy keeps the Sun radiating.

Four million tons per second may sound a great deal, but it is not much when compared with the huge total mass of the solar globe. I can assure you that you will never see the Sun shrink to a point and then pop out like a balloon!

Does the Sun have a solid surface underneath a top layer of hot gas?

No. The Sun is gaseous all the way through its globe; the temperature is far too high for any solid or liquid to form. But deep inside the Sun, the pressure is so great that the material is very dense – even though it is still a gas.

Of course we cannot see below the solar surface, but we can work out what the conditions are like below, and we know that much of the globe is made up of hydrogen. At the centre, the temperature is around fifteen million degrees celcius, and this is where the energy of the Sun is being produced (hydrogen atoms bonding together to make atoms of helium). As we have seen, the Sun is an ordinary star, and appears so splendid in our skies only because it is so close on the scale of the universe. Represent the distance between the Sun and the Earth as one inch, about the length of this line —————— and on the same scale the nearest star beyond the Sun will be over four miles away!

If I could stand on the Sun,
would I feel very heavy?

It would be rather difficult to stand on the Sun, because the surface we see is made up of gasses. You would also have to wear clothing able to tolerate a temperature of about 5,500 degrees celcius. But just suppose that you could equip yourself suitably and go on a day trip to the Sun (a journey too dangerous even for Dr Who or Darth Vader). If you weigh ten stone on Earth, you would weigh almost twenty-eight stone on the Sun. This is because the Sun is much more massive than the Earth, and has a much stronger pull of gravity.

Does the Sun spin round on its axis,
in the same way as the Earth?

The Sun certainly spins, but not quite in the same way as the Earth. Remember that the Sun is gaseous, not hard and rocky. It spins more slowly than the Earth, but the rotation period is not the same at all latitudes; it is fastest at the equator, slowest at the poles. The period at the equator is between 24 and 25 days; at latitude 40° north or south it is between 27 and 28 days, while at the poles it is as much as thirty-four days.

It is easy to check this by watching a sunspot from day to day. The spot is carried slowly across the disk as the Sun spins, and if it is at latitude fortydegrees it will pass from the preceding limb to the following limb in a fortnight (half of 27.5 days).

It will then be out of view for another fortnight, after which it will reappear at the preceding limb – unless it has disappeared in the meantime. A large spot may survive for several rotations.

Does the Sun have air like ours?

No. The Sun's 'atmosphere' is mainly hydrogen. It lies above the bright surface, which we call the 'photosphere', and with ordinary telescopes we cannot see it except when the photosphere is hidden by the Moon, during a total solar eclipse. We have to study it by using special equipment, which cuts out all radiations except those due to hydrogen. This part of the solar atmosphere is known as the chromosphere.

In the chromosphere we find 'prominences', which again have to be studied with instruments based on the principle of the spectroscope. They were first seen during total eclipses, and were, understandably, called 'Red Flames' – but they are not flames; they are made up of glowing gas. Some of them (eruptive prominences) move and change very quickly; others (quiescent prominences) may hang in the chromosphere for weeks. Well-equipped amateurs can make valuable studies of them. They are huge; the Earth is tiny by comparison.

What are the dark lines in the spectrum of the Sun, and why are they called Fraunhofer lines?

To answer this I must really go back to an experiment carried out by Isaac Newton, in 1666. He realised that sunlight is a mixture of all the colours of the rainbow, and that light is a wave motion; the colour of the light depends upon the wavelength, from red (long) to violet (short). He passed a beam of sunlight trough a glass prism, producing a spectrum; the red part of the sunbeam was bent or refracted the least and violet the most, with orange, yellow, green and blue in between. Newton never took this experiment much further, but in England in 1803 W.H. Wollaston repeated it, allowing the sunlight to pass trough a slit before entering the prism. He noticed some dark lines, but thought – wrongly – that they merely marked the boundaries between the various colours.

In 1814 a young German optical worker, Josef von Fraunhofer, again carried out the experiment. He found that the dark lines were always there, and never varied in either position or intensity; for example there were two very conspicuous lines in the yellow part of the spectrum. He did not know what caused them, but he mapped them carefully, which is why they are named after him.

The problem was solved in 1859 by two more Germans, G. Kirchhoff and R. Bunsen. An incandescent solid, liquid, or gas under high pressure will give a rainbow or 'continuous' spectrum, but an incandescent gas under low pressure will yield disconnected bright lines or an 'emission' spectrum. Each line is the trademark of one particular element or group of elements; for example sodium yields a spectrum including two very well-marked yellow lines. Some elements have highly complicated spectra; that of iron, for example, contains thousands of lines.

Now look at the solar spectrum – a continuous rainbow, crossed by lines. The rainbow is due to the Sun's surface (gas at high pressure). Above comes the solar atmosphere (gas at low pressure), which should yield bright lines. Against the rainbow these lines are 'reversed', and appear dark, but their positions and intensities are unaffected, so that they can be identified. The two dark lines in the yellow part of the solar rainbow correspond to the two bright lines produced by sodium – and this tells us that there is sodium in the Sun. By now most of the ninety-two elements occurring in nature have been tracked down in the Sun.

A spectrum of this kind is called an 'absorption' spectrum, because what is really happening is that the Fraunhofer lines absorb some of the radiation coming from below. All normal stars have spectra of the same basic type, though there are wide differences in detail.

Are sunspots really raised spots on the Sun?

No – in fact most of them are dents! They can be huge, and a really large spot can be seen with the naked eye. In 28 BC there is a Chinese record of 'a black spot as large as a coin'.

(But remember; **never** look straight at the Sun, even with the naked eye, without proper protection, such as a piece of dark welding glass). A sunspot looks black, but this is deceptive; it is about 2,000 degrees cooler than the surrounding photosphere, and appears dark by contrast. If it could be seen shining on its own, its surface brilliancy would be greater than that of an arc lamp.

But what exactly are sunspots, and how are they produced?

Sunspots are essentially magnetic phenomena. Below the solar surface run the Sun's lines of magnetic force; when these lines break through to the outer edge of the gas they cool it down – and lo and behold, there is a sunspot. In fact there will be two sunspots, sometimes quite widely separated; naturally one will be a 'north magnetic' spot white the other will be a 'south magnetic' spot. Spots usually appear in pairs, with a 'leader' and a 'follower' of opposite polarity. A group may contain many spots, and become extremely complex, but on the other hand single spots are not uncommon. No spot lasts for very long; small ones ('pores') may have lifetimes of only a few hours, though a really large group may persist for many weeks. The present record for longevity is held by a group which lasted for 200 days between June and December 1943.

A large spot consists of a dark central portion (umbra) surrounded by a tighter area (penumbra); the shape may be regular or else very irregular, and many umbrae may be contained in one penumbral mass. Drawing a complex group will tax the powers of even the most skilful artist – if you do not believe me, try it at the next possible opportunity!

Sunspots are often accompanied by faculae (Latin, 'torches') which are bright clouds hanging in the chromosphere above the spots. They may be seen in places where spots are about to appear, and may persist for some time after the spots have disappeared.

Have there been long periods when there were almost no sunspots? If so, did this affect the Earth?

Yes. The most recent of the periods of very low solar activity is known as the Maunder Minimum, because the English astronomer E.W. Maunder was one of the first to recognise it and provide the best historical analysis of it. It seems that between about 1745 and 1815 the Sun was more or less spotless; there were few displays of aurora, and when the Sun was totally eclipsed the corona was much less prominent than usual. The eleven-year solar cycle was suspended. Earlier periods of low activity have been named: the Spörer Minimum (1420–1530), the Wolf Minimum (1280–1340) and the Oort Minimum (1010–1050). It is difficult to be precise, because really good sunspot records did not begin before the nineteenth century, but as well as visual there are other methods of investigation, because prolonged solar minima affect temperatures on Earth.

During the Maunder Minimum, temperatures were well below normal; in England, the Thames froze every winter, and frost fairs were held on it. This period is known as the Little Ice Age. When the sunspots returned, there was a period of global warming; the same happened with the other minima. As trees grow, they produce rings in their trunks, and these rings tell us a great deal about the Earth's changing climates (tree ring study is known as dendrochronology). There are also definite warm periods, linked with high solar activity; for example the Medieval Maximum, and the Roman Maximum when grapes were grown in England. All in all, this seems to show that the present (2008) global warming is due almost entirely to the Sun, not to human activity.

I have heard that there was a famous astronomer who believed that there were people living inside the Sun. Surely this cannot be true?

Amazingly, it is! The astronomer was Sir William Herschel, who was born in 1738 and died in 1822. He was Hanoverian, but came to England when still a young man, and spent the rest of his life here. He was a musician, and became organist at the Octagon Chapel in the fashionable resort of Bath, but his hobby was astronomy; he made his own telescopes, and in 1781 discovered a new planet – the one we now call Uranus.

He became the greatest observer of his time, perhaps of any time, and found many new double stars, star clusters and nebulae. He was also the first to give a reasonable picture of the shape of our Galaxy.

Yet his ideas were in many respects peculiar, even by eighteenth-century standards. In his opinion most worlds supported life, and the habitability of the Moon was 'an absolute certainty', despite the fact that the Moon was known to have almost no atmosphere. The Sun, he believed, was a cool globe beneath the layer of hot gas we see, and the 'solarians' enjoyed a climate that was no more than pleasantly warm. Few other astronomers agreed with him, but his great reputation ensured that he would be taken seriously, and he never changed his views – and remember, he died less than 200 years ago; I have met a man whose grandfather actually spoke to him. The idea of a habitable Sun did not die out until years after Herschel's death.

I would be delighted to visit a Herschel-type solar base. At least the houses there would not have to install central heating!

At a total solar eclipse, the Moon just covers the Sun. The Sun is much larger than the Moon, but in our sky they appear the same size. Is there any special reason for this?

Sheer chance! The Sun's diameter is 400 times that of the Moon, but it also happens to be 400 times further away, so that the two appear virtually equal. People have tried hard to find an underlying reason, but there isn't one. The coincidence is lucky for us; otherwise we would never enjoy the spectacle of a total eclipse of the Sun. Nothing exactly the same is seen anywhere else in the Solar System. For example, Mars has two moons – Phobos and Deimos – but neither is big enough to hide the Sun, even though the Sun looks smaller from Mars than it does from the Earth (because Mars is further out). Phobos would cover less than a third of the Sun's disk, and Deimos even less; neither moon is as much as twenty miles in diameter. So unless a Martian astronomer had instruments as good as ours, he would know nothing about the solar corona and prominences – that is to say the Sun's atmosphere, which is visible with the naked eye only during totality.

On 11 August 1999 there was a total eclipse of the Sun. I went to Falmouth, in Cornwall, to observe it, but I was clouded out. Would I have been luckier if I had stayed at my home, Selsey in Sussex?

It depends what you mean. At Selsey the sky was beautifully clear, with no clouds in sight, but the eclipse was not total. Over ninety per cent of the Sun was blocked out, but all the glorious phenomena of totality – corona, prominences, Baily's Beads – could not be seen.

Eclipse stories are legion. I have always felt sorry for the man who went on a long journey to watch totality. So as not to tire his eyes beforehand, he asked a friend to blindfold him and undo the bandage just as the last segment of the Sun was covered. In the excitement of the moment, the friend forgot and all through totality the luckless astronomer sat there, unable to see anything at all. What he said at the end of the eclipse is not on record, which may be just as well.

Why are eclipses of the Sun so much less common than those of the Moon?

They aren't – there are just as many solar eclipses as lunar eclipses, but, as we have noted, the causes are quite different. A solar eclipse is due to the Moon passing into the shadow of the Earth, while a solar eclipse happens when the Moon passes in front of the Sun and hides it. Actually the name is wrong. A solar eclipse is actually an occultation of the Sun by the Moon.

The Moon's shadow is only just long enough to touch the Earth, so that the zone of totality is narrow – never more than 170 miles wide, and generally much less, so that to see totality you have to be in just the right place at just the right time. But a lunar eclipse is visible from any part of the Earth where the Moon happens to be above the horizon. This means that for any particular location on the Earth's surface, lunar eclipses really are much more frequent than those of the Sun. During the twentieth century only two total solar eclipses were visible from any part of England, those of 1927 (the Midlands) and 1999 (the West Country), but there were plenty of total lunar eclipses. Moreover, a total solar eclipse

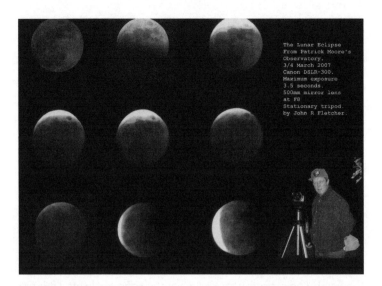

The lunar eclipse from my observatory, 3–4 March 2007. Photograph taken by John R. Fletcher

can never last for as long as eight minutes, whereas lunar totality can extend up to one hour forty-four minutes.

The last total lunar eclipse was that of 21 February 2008. The next will be on 21 December 2010 followed by two in 2011: 15 June and 10 December. The next English total solar eclipse will be that of 23 September 2090.

If I look at the Sun through my telescope, will I be able to watch the outbursts called flares?

No. Flares are violent, short-lived outbursts which generally occur above active spot-groups. They are very seldom seen with ordinary telescopes (and remember, never observe the Sun without taking full precautions), so that spectroscopic equipment is needed. The main constituent is hydrogen, so that flares show up with a filter which lets through only hydrogen light.

Flares are due to explosive releases of energy associated with magnetic phenomena. At the start of a flare event, the temperature rises very suddenly and clouds of material are sent outward through the solar atmosphere and into space; these huge 'bubbles' of material are called Coronal Mass Ejections. Radiations and electrically-charged particles are emitted, and cross the ninety-three million mile gap between the Sun and the Earth. The radiations, travelling at the speed of light (186,000 miles per second) naturally arrive first; the slower-moving particles reach us a day or two later and produce effects such as magnetic storms and displays of polar lights (aurorae). Flare activity may interfere with radio communication. Cosmic rays and energetic particles are dangerous to astronauts above the protective screen of the Earth's atmosphere, and it has even been suggested that there is a slight risk to passengers in very high-flying aircraft.

Flares are in fact remarkably powerful, and a major event may release as much energy as ten thousand million atom-bombs. Very occasionally they can be seen in 'white' light; the first observation of this kind was made by an English amateur astronomer, Richard Carrington, in 1859. However I do not recommend you go flare-hunting unless you have the correct equipment!

Does the Sun lie in the middle of the Milky Way Galaxy, and what is a cosmic year?

It was at first thought that the Sun really did mark the centre of the Galaxy, but we now know that this is not so. The real first astronomer to realise this was Harlow Shapley, in the years following the end of the First World War. He was busy studying globular clusters, which lie around the edge of the Galaxy, and found that their distribution was uneven; they were much more numerous in the southern hemisphere of the sky than in the north, and there was a marked concentration in the constellation of Sagittarius. This could be explained in only one way; we were having a lop-sided view. First measurements indicate that the Sun was 30,000 light-years away from the galactic centre, but later results have reduced this to 26,000 light-years. We lie near the edge of one of the spiral arms, the Orion Arm, between the larger Perseus and Sagittarius Arms.

The Galaxy is rotating round its centre, and the Sun's orbital velocity is 135 miles per second. It therefore takes 225 million years to make one complete circuit, and this is the period often known as the cosmic year. One cosmic year ago the most advanced creatures on Earth were amphibians; the dinosaurs

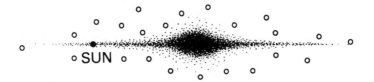

Position of the Sun in the Galaxy

came later. I wonder what life on our world will be like one cosmic year hence?

Is it possible that there is a planet on the far side of the Sun, moving in the same orbit as the Earth, so that we can never see it?

This idea of a 'counter-Earth' seems very plausible, and a planet exactly on the far side of the Sun would indeed be unobservable. This does happen when a planet passes through what is called superior conjunction; the planet, the Sun and the Earth are lined up, with the Sun in the mid position. And a planet moving in the same path as the Earth would be out of view when the alignment was perfect.

But the alignment would not last. Each planet is influenced by the gravitational pulls of other planets, and the exact lining-up of Earth, Sun and counter-Earth would be very temporary. In a matter of months counter-Earth would emerge from the glare on one side of the Sun, and we would have no difficulty in seeing it.

Will the Sun go on shining for ever?

No. As we have noted (p. 129), the Sun uses hydrogen as 'fuel', changing it into helium and releasing energy as it does so. Eventually the supply of available hydrogen will be exhausted, and the Sun will have to radiate by using other nuclear reactions.

At the moment the Sun is shining steadily, and will continue to do so for millions of years, but eventually it will become hotter and more luminous as more and more of its hydrogen fuel is used up. In a thousand million years' time, the Earth will be heated so much that life here will become impossible. Worse is to come; as the Sun ages it will swell out to become a giant star. The innermost planets, Mercury and Venus, will be destroyed; Earth may survive, but only as a molten, lifeless mass. Then the Sun will throw away its outer layers, and will shrink to become a star of the type known as a white dwarf. In the end all its light and heat will leave it, and it will become a cold, dead globe – a black dwarf.

This may sound alarming, but so far as we are concerned I promise you that there is nothing to worry about. The short-term variations in the Sun, which cause periods of global warming (as at the present time) and subsequent global cooling are very minor, and in no way significant, and the long-term changes are unbelievably slow. The Sun is the same now as it was long before the start of recorded history, and as it will be a million years hence. There will be no permanent change in your lifetime, nor in the time of your great-great-great-great-great-grandson!

Can we ever build a gun to fire supplies from the Earth to the Moon?

This was used by the great French writer, Jules Verne, a century and a half ago. He wrote a story in which the space travellers were fired Moonward in a projectile (the Columbiad) shot out from the barrel of a powerful gun, and he did his best to be accurate. In particular, he was right in saying that the travellers would have to start off at a speed of seven miles per second. This is the Earth's escape velocity, and if you begin your journey at a lesser speed you will not break free from our world's gravity.

Verne knew this, and he kept to the scientific facts as much as he could, but there is one vital point which he overlooked. Air causes resistance, and this produces heat – which is why a bicycle pump becomes warm when you pump up a tyre; the air inside the pump is being squashed. The Columbiad would have been melted even before it had left the barrel of the gun. (There was another drawback, too; the shock of departure would have turned the luckless astronauts into jelly, and in any case the journey would have been one-way only, though Jules Verne cleverly avoided this difficulty by using a wandering satellite to divert the projectile and send it back to Earth.)

Later it was suggested that the gun should be mounted on a long ramp leading to the upper atmosphere, where the air would be too thin to cause appreciable resistance, but all in all it does seem that a Verne-type gun has no future. On the other hand it might be possible to fire non-fragile payloads from the Moon to the Earth, because the Moon has virtually no atmosphere and the lunar escape velocity is a modest 1.5 miles per second. Frankly I doubt whether this will be done, but it cannot be ruled out.

Could I get into a spacecraft and fly from here to Alpha Centauri?

In theory, yes – but it would take a hopelessly long time in any of the spacecraft we use today. Alpha Centauri is over four light-years away – that is to say, over twenty-four million million miles, so that with our rockets, it is out of range. If we are to achieve interstellar travel, we must find some other method.

Doctor Who and Luke Skywalker make it sound very easy, but it isn't. If we are going in any spacecraft of the kind we can understand, we will have to work up to a fantastic speed – an appreciable fraction of the speed of light – and personally I do not believe that this will ever be done, so we must turn to what we now call science fiction. What about 'space warps', so that we can whip instantaneously from one part of the universe to another? Or 'thought travel', so that we can transport ourselves by mind power alone? Or perhaps 'teleportation', so that we can beam our bodies directly through space and re-enter them when we have reached our chosen destination?

Well – all this really is pure science fiction. But remember, television and even radio would have seemed like science fiction not so very long ago. What would Queen Elizabeth I have said if I had told her that by looking at a dial she could see men walking about on the Moon? I would probably have been arrested as a wizard. To reach the stars, we need some fundamental 'breakthrough', which may come this year, next year, in a thousand years, a million years – or never. We must wait and see. Yet we may be closer to reaching Alpha Centauri than King Canute was to *Apollo 11*.

Can there be dinosaurs on other planets?

I see no reason why not, but not in our own Solar System, because apart from Earth there is no planet suited to advanced life. I've said this before, but I'll say it again, because otherwise I won't be able to give you a proper answer.

Life on Earth began long ago, and evolved slowly. First there were primitive single-celled life-forms; these became more and more complex, so that we had sea creatures, fish and amphibians. Life spread to the land, and we had reptiles until, in what we call the Triassic period, around 230 million years ago, the dinosaurs appeared. They were very successful and dominated the scene for a long time. Then, about sixty-five million years ago, they died out, and mammals took over, ending up with you and me.

Just why the dinosaurs vanished is not known. They may just have 'worn out'; there may have been widespread fatal disease, or (as many believe) the whole climate may have been changed by the impact of a huge meteorite. But what can happen here can happen elsewhere, so look at planets of other stars. We have not seen them (yet), but we know that they exist, because of the effects they produce, and there must be vast numbers of them. With one thousand million stars in our Galaxy alone, many of them with systems of their own, one thing seems certain: there are huge numbers of planets suited to life. But as yet we have no definite proof of life anywhere except here. So this is the burning question: if life **could** appear on a planet, **will** it?

I don't know. I am sure that the answer is 'yes', but I can't prove it, and there are still people who believe that there is no life anywhere except here – just as people living many centuries ago were convinced that the Earth lay in the exact centre of the universe, with everything else rotating round it once in twenty-four hours. However, let us assume that life really is widespread, and return to the question of extraterrestrial dinosaurs.

Dinosaurs appeared here because the conditions suited them, and they remained until, for some reason or other, the whole situation changed. Now consider a world like the Earth, orbiting a star like the Sun. My instinctive feeling is that the sequence of events would be similar: primitive organisms, sea-life, fish, amphibians, reptiles, mammals, intelligent life, civilisation. But I know that this is, to put it mildly, uncertain. Life on a world slightly more massive or slightly less massive than ours might be very different in detail; why should we not have a dinosaur with twelve legs, or a mammal with three hands? We judge everything by what we see around us, and we may well be completely atypical.

So I would not rule out dinosaurs elsewhere, many light-years away, and if I could go on holiday to, say, Tau Ceti D or Delta Pavonis Q, I would be delighted to meet them. But we cannot be sure that evolving life elsewhere will go through a dinosaur stage, and it is not inconceivable that our pterodactyls and their kind really were unique to Earth.

If an alien landed here, would I be able to understand him and talk to him?

That depends on him – or her – or it. And before making any attempt to answer that question, let me repeat a number of points made earlier.

First, we cannot be sure that aliens exist anywhere. I believe they do, but as yet we have no proof of life except on Earth. Secondly, even assuming that they do exist, we have not the faintest idea of what they could be like. A being from say, Upsilon Andromedae C need not have a head, two arms and two legs. The form need not be in the slightest degree humanoid. There might not be a solid body at all. And thirdly, there might not be a spoken language. Communication might be entirely by thought transmission – telepathy – which is beyond our present power.

Clearly it might well be impossible to carry on a conversation under such circumstances, and even if our alien had a solid body and vocal chords there could still be problems. For example, it is widely believed that dolphins have a language of some kind but whether we would ever be able to learn it is another matter. Given enough time, I am confident that I could make myself understood in any terrestrial language (even Finnish!) but Dolphinspeak – well, I am decidedly doubtful.

However, there is another side to the coin. Any alien capable of crossing interstellar space will have listened in to our broadcasts, and will have learned our languages, so that our alien visitor could greet us in impeccable BBC English. This would make things very easy – but don't bank on it!

There are many wars going on at
the present time. Do you think there
may be planets on which all life has
been wiped out in this way?

With regret, I have to say that I regard this as quite possible. We
on Earth have not set a good example, and wars have become
worse and worse as we have made technological progress.
Early battles were fought with spears, and bows and arrows;
then came guns, poison gas and now atom bombs. Weapons
capable of destroying civilisation actually exist, stockpiled by the
governments of many nations. It was the Germans who first
used poison gas, and the Americans who were the first to use
nuclear weapons. Another war between leading nations would
be fatal for mankind; atom bombs would be used eventually by
whichever side appeared to be in danger of losing.

The situation has changed over the years. Originally, wars
were fought only by armies, and though millions of people were
killed there was no danger to mankind itself. During the Punic
Wars, Rome destroyed Carthage and drove a plough over the
ruins of the city. If Carthage had destroyed Rome instead, you
would still be reading this book. But if a major war breaks out
now, there will probably be no survivors.

It has been suggested that this is normal on a planet where
advanced life appears. Intelligence develops; civilisations spring
up; there are skirmishes, then a full-scale confrontation, and
finally a devastated, radioactive wasteland from which all life has
vanished. It is an appalling picture, but we cannot rule it out.

We Earthmen are currently passing through our danger period, and I always remember the wise words of Percival Lowell, who was wrong about the canals of Mars but right about almost everything else:

> War is a survival among us from savage times and affects now chiefly the boyish and unthinking element of the nation. The wisest realise that there are better ways for practising heroism and other and more certain ends of ensuring the survival of the fittest. It is something a people outgrow. But whether they consciously practice peace or not, nature in its evolution eventually practices it for them, and after enough inhabitants of a globe have killed each other off, the remainder must find it more advantageous to work together for the common good.

But Lowell wrote before the bombs on Hiroshima and Nagasaki, and it is now questionable whether there would be any survivors at all. The choice is ours. But this leads on to another problem. If civilisations are common in the Galaxy, some of them older than ours, why have they not made contact with us? In fact, as one eminent astronomer asked, 'Where is everybody?'

There are several possible answers. The first is that there is nobody there at all, and this is still the belief of some people, though not many. The second is that the distances are so vast that there is simply no way of bridging them, and no civilisation can do so. The third is that there are civilisations quite capable of getting in touch with us, and choose not to. It is quite true that one culture faced with another which is more technologically advanced may suffer; one has only to look at what has happened to the American Indians or the Australian Aborigines. Though other races may be able to travel between the stars, it is conceivable that they may feel that we are best avoided. After all, some of our own astronomers have expressed doubts about the wisdom of trying to contact 'aliens'. I suppose it all depends on one's point of view.

We are sending radio messages into space, hoping that someone will answer. Some astronomers call this dangerous. Do you agree?

No, but I can see why some people believe that I am wrong. Let me give you my reason.

We can be absolutely certain that there is no intelligent life in our Solar System except (possibly!) on the Earth. This means that the nearest intelligent race must be several light-years away, totally beyond the range of our rockets. We can reach the Moon and the planets, but we have absolutely no idea of how to cross interstellar space. Neither can we be one hundred per cent certain that other civilisations exist anywhere, though there are millions upon millions of other Solar Systems, and it seems both conceited and illogical to suggest that life is unique to Earth – the third planet in the family of a very ordinary star.

If there are other races, they may differ from you and me both mentally and physically. They might be far more advanced technologically than we are, and this is why there have been misgivings about trying to contact them. We know, sadly, that when two peoples meet, war often follows; we have only to look at what has happened here – the Native Americans were almost wiped out by the white invaders; the Australian Aborigines suffered similarly, and so did the African Bushmen. It has been claimed that a super-intelligent race from across the Galaxy could come here and either subjugate us or destroy us. So the policy should be: 'Keep quiet, and hope that nobody will realise that we are here'.

But think about this more carefully. If any 'aliens' are capable of reaching us, they must be enormously more advanced than we are, and are not likely to come to us with conquest in mind. It is reasonable to believe that any civilisation will pass through a stage where it knows enough, technologically, to destroy itself. We are in that stage now, with enough stockpiled nuclear bombs to turn the whole Earth into a dead, radioactive waste. Whether we will actually do so, I hope not; if we come to our senses, we can progress, and eventually, no doubt, find out how to reach the stars. You can probably see where my thoughts are taking me. A civilisation able to cross interstellar space will have passed through its 'danger period', and will come in friendship – otherwise it would not have survived.

Children learn about science from a very early age – astronomy in particular. But when did it begin, and who were the first great astronomers?

Men must have looked at the sky from the Stone Age and before, but there was no science of the kind we know today – and neither would there be. Understanding was slow to develop. The Chinese kept records of celestial phenomena, such as comets and eclipses, and even had some idea of how to predict eclipses (there is a story, certainly pocryphal, that in 2136 BC the two Court Astrologers, who rejoiced in the names of Hsi and Ho, were executed for failing to give due warning that an eclipse was due – remember that the Chinese believed in a hungry dragon which was trying to gobble up the Sun – much displeasing Emperor Hung Kang). The Chinese had their constellation patterns, very different from those we use today, and they noted strange events such as 'guest stars', or supernovae. But that was about as far as they went, while the Egyptians believed that the sky was formed by the arched body of the goddess Nut. Of course, the Egyptians too add their constellations – different from the Chinese, different from ours.

True science began in Greece. Many of you here will know vastly more about Greek history than I do, so what I propose is to give some of my own personal comments, with the full realisation that many people will disagree with them. And as a start, we must surely remember that Greek science did not rise overnight. Thales, first of the great philosophers, flourished about 600 BC. Ptolemy, the last, died about AD 180. The two were separated by almost 500 years, so that in time, Ptolemy was as remote from Thales as we are from the Crusaders.

Of course, our records are very fragmentary. Thanks mainly to Ptolemy, we know a great deal about ancient science, but there is much that has been left out, and in few cases do we have any idea of the careers or even the personalities of those who were involved. My main theme is this. The Greeks made many discoveries, and they broke down many scientific barriers. Indeed, some of them were aware of the status of the Earth, thereby anticipating Copernicus by nearly two thousand years. In other ways their ideas sound strange today; but – and this is my point – how could they have done any better?

Thales, first in our list, believed that the Earth was flat, and floated on water. Well, take a look around you. Allowing for local irregularities such as hills and valleys, the Earth really does look flat. We know that it is a sphere, because we were told so from childhood; but nobody had told Thales, and he had to decide for himself. At least he looked upward and observed the sky; he had something upon which to build. He could have read Homer, to whom the sky was a solid vault straddling the Earth, with the fiery gleaming aether above the cloudy air!

A cylindrical Earth as with Anaximander, around 550 BC, with celestial bodies shaped like 'fiery wheels'? Light from them came out in jets; and an eclipse was due to the partial closure of a jet. With Anaximenes (c. 525 BC) a flat Earth floating on air, or Heraclitus, whose Sun, Moon and stars were fiery bowls; the Moon, relatively close to us, travels through the less pure lower air, while the Sun, higher up, is 'the closest and hottest of the stars' – a new concept, the stars are suns.

With Pythagoras we come to a certain amount of mysticism; the central object is the Hearth of the Universe, while the Sun is a glass sphere reflecting the 'hearth light', and the supreme Hearth itself is hidden from us by an intervening body, a counter-Earth. Impure Man must not be allowed to see the most important of all bodies. And Empedocles (mid-fifth century BC) with his complex universe; an outer hard shell on which the stars are fixed, with an inner shell of double hemispheres, one of fire for day and the other dark for night. But now the Sun and Moon are merely concentrated polished spots on the inner surface, which reflect the outer fire.

Varied ideas? Certainly; but gradually things changed. One important point is that persecution on 'religious' grounds was very muted. True, Anaxagoras was banished from Athens for teaching that the Sun is a red-hot stone larger than the Peloponnesus, but this is very mild compared with Christian persecution so much later; recall in 1600 Giordano Bruno was burned at the stake in Rome for his 'crimes' (he believed the Earth to be in orbit round the Sun). In 1633 Galileo was condemned by the Inquisition for the same offence. He was finally exonerated. I agree, and the Church admitted that he had been right: but it was not until 1992! Nobody can therefore accuse the Vatican of making hasty decisions.

Aristotle (384–322 BC) knew that the Earth is a globe; for one thing, the Earth's shadow on the Moon during a lunar eclipse is curved, so that the Earth's surface must also be curved. Also, the altitude of the Pole Star changes according to the latitude of the observer. Then came Aristarchus of Samos, who lived from around 310–250 BC. I must pause here, because Aristarchus seems to me to be a key figure.

We do not have his original text, but we do have the words of Archimedes:

> Aristarchus of Samos brought out a book consisting of certain hypotheses, in which his premises lead to the conclusion that the universe is many times greater than it is presently thought to be. His hypotheses are that the fixed stars and the Sun remain motionless and that the Earth revolves about the Sun in the circumference of a circle, the Sun lying in the middle of the orbit, and that the sphere of the fixed stars, situated about the same centre as the Sun, is so great that the circular orbit of the Earth is as small as a point compared with that surface.

Aristarchus made a gallant effort to measure some of these quantities. He attempted to measure the angular separation of the Sun and Moon at the time when the Moon is at exact half-phase (dichotomy), since this would be key. His results were inaccurate, because he gave the angle of the Sun as 87° (it is in reality 89°50'). The lunar surface is so jagged that there is no chance of making a good timing of dichotomy. But – and this is important – his method was absolutely correct in theory, and it was not his fault that he lacked the means to make precise measurement. The same is true of his various other estimates.

He concluded that the distance of the Sun from the Earth is between eighteen and twenty times greater than the distance of the Moon, and that the Sun is a great deal larger than the Earth.

Why did not Aristarchus' ideas take root? Quite simply, because they could not be proved – and old traditions die hard. It was more of a question of technological limitation than of prejudice. It was different with Eratosthenes' measurement of the circumference of the Earth, around 240 BC. When the Sun was overhead from Alexandria, and shone straight down a well, it was not overhead at Syene (the modern Aswan). By measuring the distances and the angles, Eratosthenes gave a very accurate result. But these were measurements which were definite, straightforward and could be repeated. Those of Aristarchus were not.

I will not labour these points; I hope I have given a sufficient number of examples. Around 140 BC Hipparchus of Rhodes developed really accurate methods of measurement, and drew up the first good star catalogue; indeed, comparing his results with those of earlier works, he discovered the precession of the equinoxes. But we come finally to Ptolemy, whose work marked the end of the era of classical astronomy.

Ptolemy lived in Alexandria, but must surely have been Greek. It is a great pity that we know nothing about his career, or even what he looked like; I have often wondered! It is also sad that his great book has come down to us only by way of its Arab translation. At least we must be grateful for small mercies. It might so easily have been lost altogether.

.

Ptolemy's interests ranged far and wide; in particular he drew up the first map of the then-civilised world which was based on scientific measurement rather than guesswork, and it was fairly good; Britain is shown, though Scotland is fastened onto England in a sort of back-to-front position. However, I would challenge anyone to do better with the resources available to Ptolemy. Probably nobody else could have done so well.

He drew up a star catalogue, which was based on Hipparchus', but with large input from Ptolemy himself, and he measured the movements of the planets against the starry background. Finally he gave the theory which to us is always known as the Ptolemaic; true, Ptolemy did not invent it, but he brought it to its highest degree of perfection, and he was of course a master mathematician.

You will all know the Ptolemaic theory. The Earth is central, spherical but non-rotating; round it moves the Moon, Sun and planets, in circular orbits – the circle is the perfect form, and nothing short of perfection can be allowed in the heavens; beyond comes the sphere of the fixed stars. The movements of the planets cannot be accounted for by assuming that they move round the Earth in circular orbits at steady speed; Ptolemy had to use various devices – eccentrics, quantums and above all epicycles – in order to make the system work. But in the end, work it did. It fitted all the facts as Ptolemy knew them just as well as Aristarchus' heliocentric theory had done, and at the time it no doubt sounded a great deal more logical.

Occasional attempts to belittle Ptolemy, and to claim that he was at best a copyist who was not above fudging his observations, have been originally unsuccessful. He deserves his nickname of 'the Prince of Astronomers', and without him we would know much less about ancient science than we actually do. With his death came the start of the Dark Ages, and it was many centuries before others began to take up where Ptolemy had left off.

The medieval era had a head start; there was the Greek work to act as a basis – and the Greeks themselves had never had anything comparable. They were logical, and they were always anxious to 'save the phenomena' – in fact, to make theory and observation agree. In the end they were successful, and looking back at what they achieved I maintain that there was no way in which they could have done any better.

My case rests.

Did you ever know Sir Arthur Clarke, the man who predicted communications satellites?

I first met Arthur Clarke in 1935, at a meeting of the British Interplanetary Society (BIS) – now a much-respected organisation, but in those days generally classed with Flat Earthers, Creationists, and people who believe that Bacon wrote Shakespeare. He was then seventeen, and already starting to make his mark, while I was a timid newcomer of twelve. He came across the room to greet me, and despite the difference in our ages we struck up an immediate friendship which proved to be life-long.

We saw each other regularly at meetings, but then came the war; the BIS went into a state of hibernation, and Arthur and I went our different ways. We both joined the RAF. He became a radar expert, while I flew as a navigator – I think we may have been commissioned on the same day in 1940 – but our paths did not cross again until Herr Hitler and his unsavoury colleagues had been removed. The BIS re-started; Arthur became President, while I was elected to Council and became the first editor of the periodical *Spaceflight*. The BIS still had problems, and it is on record that Arthur, known to be an enthusiastic member, was once stopped and closely questioned when going into the Science Museum, because the doorman thought that he might be carrying a bomb.

Council meetings were serious and constructive, but there were lighter moments too. The 'flying saucer' craze started, and we had some curious letters. One, I well remember, was from a particularly earnest lady, and began as follows: 'I have proof that Flying Saucers come from the planet Strorp, 200 light-years away'. How to reply? We thought very carefully, and finally settled on an answer: 'Many thanks for your letter. However, we must tell you that you have been misinformed. The Saucers do not come from Strorp, but from the much more distant planet Ploop, 400 light-years away.' We heard no more!

Both of us wrote, and in Arthur's case with spectacular success. In his famous article in *Wireless World* he invented communications satellites (!) and he made many other predictions, many of which have come true. I remember one discussion I had with him – I'm not quite sure when, somewhere around 1953, I think. We were talking about travel to the Moon, and Arthur said that in his opinion men would go there before 1970. I gave him a pitying smile. '1970? Not a chance. Not before 1980 at the earliest, and we'll be lucky to do it before 1990.' He reminded me of that when Neil and Buzz stepped out on to the Sea of Tranquillity. Mind you, he wasn't always right. He was not as profoundly sceptical about flying saucers as I was, and he was sure that by 2000 we would have made contact with an alien civilisation. I wasn't. We laid a bet on this, and I won. Arthur duly paid up (one bottle of white wine).

It is just possible that I made one tiny contribution to the classic film *2001: A Space Odyssey*. We were having lunch together and chatting about suitable music. I am a Viennese waltz fanatic, and for the space station sequence I said I'd go for the *Blue Danube*. They did. Whether my casual remark had anything to do with this, I don't know. Probably not, but I do remember that comment!

I also remember another mildly amusing episode. On one occasion we both wrote books for young readers. They didn't clash, but his had a foreword by me and mine had a foreword by him. Coincidentally, and to the surprise of both of us, they were published on the same day. Not that it mattered, because his was essentially astronautical while mine was purely astronomical.

Arthur was always keen on underwater exploration, and Sri Lanka was a good site; he went there for a holiday, liked it, and made his home there. I had an open invitation, but every time we made an arrangement something went wrong, and in the end I never made it. I am also sorry not to have played Arthur

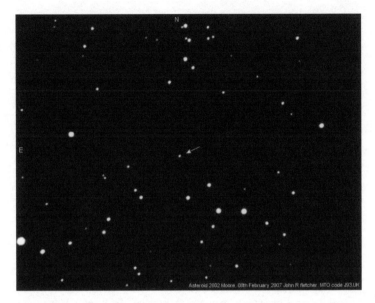

Asteroid 2602 Moore (arrowed). Photograph by John Fletcher (the arrow was added later!)

at table tennis – one of his favourite hobbies. He was a good defensive player and decades ago I attained county status – it would have been quite a battle.

Though I never got to Sri Lanka, we were always in touch by letter, phone and television; the last time we met face to face was when he came over to receive his knighthood. Then, of course, came the tsunami which caused such havoc. Sri Lanka was right in the track; fortunately Arthur's home was undamaged, but many people were left homeless. A thought came to me. Somehow or other the two of us had never produced a book under joint authorship; why not do so now, and give the proceeds to the island's disaster fund? My suggested title was *Asteroids*, because each of us had had one of these tiny bodies named after us (Arthur commented that he wasn't sure which of us was the larger or the more eccentric). We went to work at full speed. David Hardy provided lovely illustrations, the Astronomer Royal wrote a foreword and the *Daily Express* printed and bound it – all without charging a penny. In the end I believe that that little book raised around £30,000.

In his last years Arthur's health failed, but his brain remained as clear as ever. I spoke to him on the phone only a week before he died, and he was still making plans for the future. He will always be remembered as one of the great visionaries of modern times, and I am proud to have known him so well and for so long.

What happened to the first Russian space station, *Mir?*

It was deliberately de-orbited and burned away while descending through the atmosphere – a sad end to a great project. During its final years it had a bad press – 'Russia's flying scrapyard', and so on – but one important point must be remembered. It was put into orbit, from the Baikonur cosmodrome in what was then the Soviet Union, on 20 February 1986. It was scheduled to remain active for no more than five years. In fact it remained aloft for more than a decade and a half, and almost all the really serious problems occurred well after it had passed its original planned lifetime. For the first five years of the mission it achieved almost everything that had been expected of it.

It had various functions. First, it was a scientific vehicle designed to carry out investigations which are difficult or even nigh-impossible from ground level; we have to peer through a thick layer of atmosphere, whereas Mir had a clear view from an average height of almost 250 miles above sea level, right on the fringe of space. Secondly, it provided information about the way that the body of a human being would adapt to conditions in space, under conditions of weightlessness. And thirdly it developed into a truly international venture. Of course it was Soviet-built and Soviet-launched, but before long astronauts of many other nations would go to it: Americans, Britons, Arabs, Afghans, Japanese – all were able to take part. They spent sometimes prolonged periods on Mir and many of them played very important roles. In almost all cases the relations between the astronauts were good – which, considering that all of them had been hand-picked, is not surprising.

The first section of Mir, known as the core module, was first crewed by two Soviet cosmonauts, Leonid Kizim and Vladimir Solovyov; in all Solovyov made five trips to the station. Later further modules were added: Kvants 1 and 2 (mainly astronomical); Kristall (medical and biological); Priroda (studies of the Earth's atmosphere); and Spektr (studies of particles in low Earth orbit). Spektr was the last of these to be added, well after the end of its originally planned lifetime; but even so, most of the equipment was still working well. Only in the later stages were there problems which were undoubtedly life-threatening. The worst of these was a collision between Mir and an unmanned cargo vehicle, one of the *Progress* series on 2 June 1977, when the space station was past its heyday.

This was quite unexpected. The cargo vehicle had been filled with unwanted rubbish and detached from *Mir*. Normally it would have been allowed to drop back and burn away in the atmosphere), but on this occasion it had been decided to practice the docking technique which had caused some trouble in the past. On board *Mir* were two Russians, Viktor Tsibliyev and Alexander Lazutkin, together with Michael Foale, the first British astronaut to undertake a 'space walk'. Foale was born in Lincolnshire, of an English father and an American mother; he took his degree in astrophysics at Cambridge University and then moved to the United States to join NASA. His first comment during his initial space walk was 'Golly it's high up here!' which may not have been as dramatic as Neil Armstrong's 'one small step', but was at least unscripted!

The *Progress* vehicle was of course remote-controlled, and Tsibliyev was in charge of it. The procedure was to bring it gently in to the docking port, but it has been said that it was rather like backing a heavy lorry into a smaller car park, and on this occasion Tsibliyev miscalculated. *Progress*, travelling much too fast, headed straight for *Mir* and struck the Spektr module, puncturing it and destroying one of the solar panels which provided *Mir* with its power. The danger was very real. As soon as Spektr was struck air began to hiss out, and the pressure started to drop. Michael Foale and Lazutkin had to act quickly. They managed to cut the cables connecting Spektr with the rest of the station and then slam and shield the door. Meanwhile

Tsibliyev was calling Mission Control in Moscow, because flight rules laid down that in an emergency of this sort the crew should evacuate *Mir* and escape in the Soyuz capsule which was attached to the station and was to be used for the eventual return to Earth. Mission Control ordered the astronauts aboard, and in the end the damage did not prove to be serious but it had been a frightening moment, as Michael told me when I talked to him a few months ago. His quick thinking had been decisive. Tsibliyev said later: 'He helped us more than many American astronauts'.

There was another crisis in 1997, when a canister filled with inflammable material caught fire and started a blaze which lasted for a minute and a half. Dense smoke filled the living and work quarters of *Mir*, and the astronauts – two Russians, one German and one American, Jerry Linenger – had to put on oxygen packs and breathing masks while the automatic control system cleansed the station. The blaze was put out by four fire extinguishers. In a way it may have been beneficial; it was said that by this time the smell inside *Mir* was like that of a very old garage which had not been washed for a long time.

One cosmonaut, Segei Krikalev, made two trips to *Mir*. On the second occasion he met with a curious situation. On 18 May 1991 he and his companion, Anatoli Antsebarsky, took off from Baikonur and arrived on *Mir*; both were of course members of the Communist Party of the USSR. They were due back in August, but around this time the Soviet Union disintegrated, the mission's landing site was obtained by Kazakhstan, and the return was postponed. A Kazakh cosmonaut was sent to replace Artsebansky, but Krikalev was left on *Mir*, orbiting at

250 miles at 17,000 mph and reduced to the status of a helpless spectator.'I have a question' he said to Mission Control.'Are you going to sell the space station – along with me?'When he finally landed at Kazakhstan in March 1992, he had been aloft for 310 days and the USSR was no more; as a Russian, perhaps Krikalev was lucky not to have been asked for an entry visa. He made another trip to *Mir* later (March 2001).

Krikalev, like many space men, is an enthusiastic athlete (Michael Foale is another) but in orbit there is no 'weight' and it was once thought that living under zero gravity for more than a few days would be dangerous. In fact, there are cosmonauts who have been on *Mir* for over a year, and these experiments have been of the utmost value to medical researchers. If we are ever to go to Mars, we must be sure that the human body can cope. Not long ago I discussed this with Musa Manarov, who spent a total of 541 days on *Mir*; he made two trips, of which the first lasted for exactly a year.'My condition remained good' he told me, 'of course I took all precautions, including regular exercises.' It would be wrong to say that weightlessness is not harmful, in the long run, and much research is yet to be carried out, but *Mir* has made a vital contribution.

Scientific work was carried out all the time that *Mir* was inhabited. Physics, astronomy, botany, environmental studies, meteorology … all have been to the fore, and all in all it must be said that despite its later problems, *Mir* was a triumph. It has been an essential step in space research; without it the *International Space Station* could never have been launched. Sadly in August 1999 it came to the end of its active career, and was abandoned. A brief trip in April 2000 was the last time it was ever visited.

I had hoped that *Mir* would be boosted into a higher, safe orbit and left there until it could either be renovated or else fished down, but it was not to be. I saw it during one of its last orbits, and I could not help feeling rather sad. On 23 March 2001 it came to the end of its career, and what was left of it fell into the south Pacific Ocean. Today we have the *International Space Station* – but it was the achievements of *Mir* that made the *ISS* possible.

When will Earthmen be able to travel to other Solar Systems?

I wish I could answer that question, but frankly I can't. All I can do is to make some general comments.

One thing is fairly obvious: modern-type rockets are useless, because of the sheer distances involved. We can reach the Sun's planets, and we have done so; we are dealing with millions of miles, and this is manageable. Mars can be reached in a few months, and so can Venus (not that anybody is likely to want to visit Venus). *Voyager 2* took twelve years to get to Neptune, and if men do want to go there it is possible. But all the stars are millions of millions of miles away, and this means that we are facing a totally different situation.

A journey in even our fastest rocket would take many centuries, and we can rule out conventional journeys. Science fiction writers have not been short of ideas, and of these two have been particularly popular. Could the travellers be put into a state of suspended animation, remaining inactive until reaching their destination and waking up? I very much doubt whether human bodies would survive, and in any case – who would wake them up? We do not know the location of the nearest planetary system, or where we could find a world where people such as ourselves could live. No, I believe that this method of interstellar travel must be left to the science fiction writers.

Next, the Space Ark. A party of astronauts sets out, knowing full well that they will die long before arrival. Both male and female pioneers take part; children are born, grow up and die; generation follows generation, and only the descendants of the original pioneers survive to make landfall – or, rather, planetfall. One can imagine the comments as the Ark touches down on Epsilon Eridani Q after a journey of more than ten centuries. 'I wonder if this is anything like the planet Great-Great-Great-Grandfather describes in that recording he made for the Library – Earth, he calls it, with seas and trees, and a blue sky? There was rain, too; water pouring down through the air and making everything wet and cold.' This is, surely, almost as far-fetched as the suspended animation picture.

Well, can we make spacecraft capable of travelling faster than light? According to Einstein's theory of relativity, the answer is 'no', because at speeds of that kind very strange things start to happen. Your mass increases, and your time slows down; at the full speed of light your mass becomes infinite and your time stands still – which is another way of saying that it can't be done. Up to now Einstein's theories have passed every test made, and it seems that we can discount faster-than-light rockets.

If I am right about all this, we cannot reach other planetary systems by means of material spacecraft, and we must look elsewhere. Space warps, for example. The space concept is not so straightforward as it looks; could we find a 'short cut'? Consider a river meandering through the countryside. You want to reach a field opposite; by following the bends in the river (if for example you are in a boat) it will take some time, but if you are sufficiently athletic you can jump straight across. You are using a short cut, but to find a short cut across space is not nearly so easy.

Teleportation, the art of leaping from one point to another in no time at all? Or thought-travel, when you make the journey without bothering to take your body with you? Now we really are deep in the realm of science fiction, and we have to admit that we have no idea of how this sort of thing can be done, even if it can be done at all. But there is another relevant point to be made here.

Think back to the time when Queen Victoria came to the throne in 1837. Think what the average person would have said about pushing a button, looking at a screen, and watching men walking on the surface of the Moon. 'Ridiculous! Utterly absurd. It can never happen.' Yet within a century and half, it did. Science fiction has a habit of turning into science fact, and to our average Victorian, television was as unreal as teleportation is to us. So I am not saying that interstellar travel is impossible; I am merely saying that it is impossible by any means known to us at present, which is certainly not the same thing.

Our civilisation is only a few thousand years old, and our technological ability is much younger than that. Other civilisations, in other systems, may well be older and more advanced; why not? They could have mastered interstellar travel, in which case they could be capable of visiting Earth if they chose to do so. Unfortunately, any serious discussion here is bedevilled by the 'giggle factor' – flying saucers, alien abductions, etc, which are pure pantomime. But it is not impossible for an interstellar spacecraft to land here; it has not happened yet, at least in historic times, but one day it may. I hope it does. Visitors from afar will surely have left us well behind, both technologically and culturally, and would be able to teach us a great deal. Any race capable of crossing 'deep space' will have become too adult to indulge in warfare, as, alas, we of the twenty-first century still do.

Have there been any astronomers who have been universally disliked?

There have been one or two. It was said of the nineteenth-century French astronomer Le Monnier that he never failed to quarrel with anyone he met – and he was not exactly systematic; one of his pre-discovery observations of the planet Neptune was scrawled on a bag which had contained hair perfume. Urbain Le Verrier, whose mathematical calculations did lead to the discovery of Neptune in 1846, was also unpopular; one of his contemporaries said that although he might not be the most detestable man in France, he was certainly the most detested. There was also Fritz Zwicky, who described his colleagues as 'spherical bastards'. But perhaps the palm must go to the American astronomer Thomas John Jefferson See (T.J.J. See) who was, at one time around a hundred years ago, on the staff of the Lowell Observatory at Flagstaff, in Arizona. He was due to depart, and another Lowell astronomer, A.E. Douglass, was not sorry:

> I have never had such aversion to a man or beast or reptile or anything disgusting as I have had to him. The moment he leaves town will be one of vast and intense relief, and I never want to see him again. If he ever comes back, I will have him kicked out of town.

Fortunately, See did not return to Flagstaff. At least the episode shows that astronomers are human!

Do all stars have their own names?

To give each star a name would be rather difficult – how do you think up millions of names? The very brightest stars do have individual names – Sirius, Canopus, Rigel and so on – but the proper names of fainter stars are not generally used except in a few special cases, such as Polaris (the Pole Star), Mizar (the double star in the Great Bear) and Algol (the eclipsing binary in Perseus).

A different system was introduced in 1704 by a German amateur astronomer named Bayer. He used the letters of the

Greek alphabet, beginning with the first (Alpha) and working through to the last (Omega). The brightest star in each constellation was lettered Alpha, the second brightest Beta, the third Gamma, and so on. Thus Mirphak, the brightest star in Perseus, became Alpha Persei; Algol, the second brightest, became Beta Persei and so on.

It sounds all right, and basically it is; we still use it. However, the letters are often out of order, and in Orion, for instance, Beta (Rigel) is brighter than Alpha (Betelgeux), while in Sagttarius (the Archer), the brightest stars are Epsilon and Sigma; both Alpha and Beta are much fainter. In Octans, the south polar constellation, the brightest star is Nu. Some constellations have incomplete sequences; in Leo Minor (the Little Lion) there is no Alpha, and only one of its stars, Beta, has a Greek letter. Some stars have been transferred from one constellation to another; thus Alnath, which used to be Gamma Aurigae, has become Beta Tauri, while Gamma Scorpii has become Sigma Librae.

I was born on 4 March, and I am told that my ruling star sign is Pisces. What does this mean, and is it important?

It is not in the least important, and it means nothing at all. This is astrology, not astronomy, and there is no connection between the two. Astrology claims that there is a link between the positions of the planets and human character and destiny, but a moment's thought will show how silly this is.

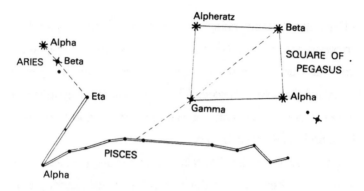

Pisces (the Fish), near the Square of Pegasus. Also near is Aries (the Ram)

The stars have been grouped into constellations; Orion, the Great Bear and so on. But the stars are light-years away, and the stars in any particular constellation have no real connection with each other; we are dealing simply with line of sight effects. Watch a low-flying aircraft with a high cloud in the background. There is no connection between the two. More importantly, we can give what names we like. We happen to follow the patterns drawn up by the Greeks, but these were quite different from the constellation names given by other early stargazers, notably the Chinese and the Egyptians. In fact, what we call a 'constellation' is simply a name that we have given to stars which lie in roughly the same direction as seen from Earth.

Next, the planets. They are much closer than the stars; if one represents the Earth-Sun distance by one inch, even the nearest star will still be over four miles away. As I write these words, Saturn is seen against the background of the stars in Leo (the Lion), but to say that Saturn is 'in' Leo is as senseless as saying that an aircraft is 'in' the cloud.

The Solar System was formed from a disk of material surrounding the young Sun (we are now looking back about four and a half thousand million years in time). Because this disk was flattish, the Sun, Moon and planets keep to a definite belt round the sky, known as the Zodiac. The Greeks divided the Zodiac up into twelve constellations, some of which were large and bright while others were smaller and obscure. The twelve constellations are: Aries, the Ram; Taurus, the Bull; Gemini, the Twins; Cancer, the Crab; Leo, the Lion; Virgo, the Virgin; Libra the Scales; Scorpus, the Scorpion; Sagittarius, the Archer; Capricornus, the Sea-goat; Aquarius, the Water-Bearer and Pisces, the Fishes.

For some reason or other astrologers always refer to the Scorpion as 'Scorpio'. Only one name does not refer to a living creature, but Libra was originally Chelae Scorpii, the Scorpion's Claws. When this Zodiac was originally drawn up the Sun was 'in' Aries at the beginning of the year, so that the Ram became the first of the Zodiacal constellations.

From the astrologers' point of view there are several immediate problems. First, an extra constellation, Ophiuchus (the Serpent-bearer) crosses the ecliptic – the middle line of the Zodiac – for some way between Scorpius and Sagittarius, so that the Sun, Moon and planets move into it regularly, and there are in fact thirteen Zodiacal constellations, not twelve. Secondly, the Sun is no longer in Aries at the beginning of the year; it now crosses the ecliptic in the adjacent constellation of Pisces, so that the astrological 'signs' are out of step with the constellations after which they are named. Finally, yet another constellation, Cetus (the Whale) skirts the Zodiac, and planets may enter it briefly.

Astrologers are not perturbed by these awkward facts, though they have an intense dislike of Ophiuchus (they can gloss over Cetus). Astrology still has many followers, but it is best reserved for seaside piers, circus tents, and columns in tabloid newspapers. I can cite one personal experience here. One a summer day I read my astrological forecast, and learned that I was due to achieve an outstanding athletic performance; I was playing cricket, and on a spinner's wicket, which should have suited me, I took 0 for 62. Clearly my stars were out of touch.

Why are the stars divided into constellations, and who named them?

This is a question which needs to be answered in full, because it raises so many points.

First, the names of the constellations are quite arbitrary. We happen to base our naming system on those of the Greek astronomers of over two thousand years ago, but there were earlier systems too, notably by the Chinese and the Egyptians. If we followed, say, the Egyptian system our star maps would seem very different; there would be no Lion and no Crab, but there would be a Cat and a Hippopotamus, though of course the stars themselves would be exactly the same. Really, you can give whatever names you like, because the constellation patterns have no true significance.

The stars are at very different distances from us, and what we call a 'constellation' is nothing more than a line of sight effect.

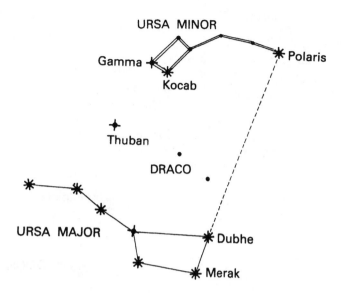

The Bears (Ursa Major and Ursa Minor), with the Pole Star

Consider the bright stars Castor and Pollux, in the constellation of Gemini, the Twins. They lie side by side in the sky, but there is no real connection between them; Pollux is thirty-three light-years away from us according to a very recent estimate, while Castor's distance from us is fifty-three light-years. Castor is 'in the background', so to speak, and its actual distance from Pollux is over twenty light-years i.e. around 120 million million miles. If we were looking at them from a different vantage point in our star system or Galaxy, Castor and Pollux could well be on opposite sides of the sky.

The last of the great astronomers of Classical times was Ptolemy, who lived from around AD 120–180 (we are not too sure about his precise dates). He gave a list of forty-eight constellations, and

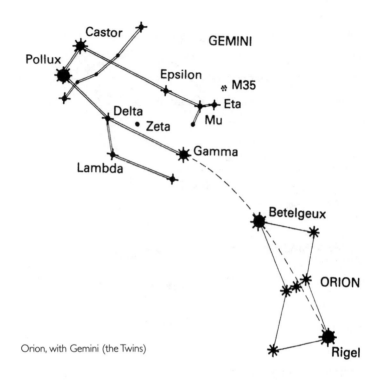

Orion, with Gemini (the Twins)

all these have been kept on our maps, but in most cases the outlines have been modified. Also, Ptolemy lived in Greece, where the southernmost stars never rise, so that these had to be given constellation names by later observers. Today the International Astronomical Union, the controlling body of the world of astronomy, gives a list of eighty-eight constellations. It is certainly not ideal, but we have become so used to it that it will not be altered, and everybody accepts it.

The names are used in their Latin form; after all, Latin is still the international language, even though it is no longer spoken in ordinary conversation. Thus the Great Bear becomes 'Ursa Major', the Lion is 'Leo', the Scorpion is 'Scorpius', and so on.

Some of the added constellations have decidedly modern names – such as Microscopium (the Microscope) and Antlia (the Air-pump). There are some mythological characters, notably Orion and Cassiopeia, and it has been said that the whole sky has been made into a storybook.

However, the constellations are very unequal in size and brilliance, and a famous nineteenth-century astronomer, Sir John Herschel, once commented that they seemed to have been designed so as to cause as much confusion and inconvenience as possible. Some of them, even a few of Ptolemy's, are so obscure that they do not seem to merit separate existence. On the other hand, one original constellation, Argo Navis (the Ship) was so large and unwieldy that eventually the IAU lost patience with it, and chopped it up into a keel, sails and a poop!

Could I make up a constellation, and name it?

There is no reason why you shouldn't – but nobody would take the slightest notice, and it would be a complete waste of time.

Over the years, various astronomers have proposed to form new constellations made up from stars of existing groups but when the International Astronomical Union finalised the list, most of these proposed groups were rejected. Out went Sceptrum Brandenburgicum (the Sceptre of Brandenburg), Honores Frederici (the Honours of Frederick), Officina Typographi (the Printing Press), Lochium Funis (the Log Line) and others. But one, Quadrans Muris (the Mural Quadrant), formed in 1775 by the famous astronomer Johan Elert Bode from stars in the large constellation Boötes (the Herdsman), is still remembered, in a way. Every January, meteors are seen radiating from that area, and we still call these meteors the Quadrantids.

I rather regret two more rejected constellations, Noctua (the Night Owl) and Felis (the Cat). Why shouldn't the sky have the owl and the pussycat?

There are also people who claim to sell new star names. Various agencies say that they can do this, naturally, for money. 'Send us £50 and we'll name a star after your daughter.' Frankly, this is an unpleasant confidence trick, made worse by the fact that the advertisements are cleverly done and look convincing, but the 'names' mean nothing and the money is wasted. Have nothing to do with schemes of this sort. I am reminded of the naïve American tourist who agreed to pay a salesman several thousand for the privilege of re-naming the Eiffel Tower. I do not know whether the French authorities heard about this – if they did, they will not have been very impressed!

Which are the largest and smallest constellations in the sky? And why aren't they all the same size?

The largest is Hydra, the Watersnake, which covers 1,303 square degrees – and has only one star, Alphard, above the third magnitude. The only other constellations with over 1,000 square degrees are

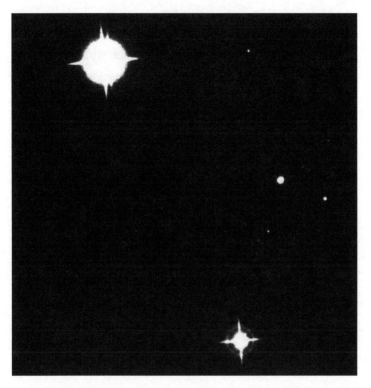

Alcor and Mizar. Ursa Major. Photograph taken by John Fletcher

Centaurus (1,060), Cetus (1,232), Eridanus (1,138), Hercules (1,225), Pegasus (1,121), Ursa Major (1,280) and Virgo (1,294). Argo Navis used to be much larger than any of these, and was so unwieldy that it has been chopped up into Carina, Vela and Puppis.

Surprisingly, the smallest constellation is Crux Australis, the Southern Cross, with only sixty-eight square degrees, but it has two stars of the first magnitude and another only just below. Also below 100 square degrees are Circinus (93), Equuleus (72) and Sagitta (80).

The constellation patterns we use are Greek – and the Greeks turned the sky into what is to all intents a storybook, without caring in the least about orderliness! Some constellations added later are very obscure; one, Mensa, has no star above the fifth magnitude.

Do any of the constellations look anything like the objects after which they are named?

Very few. I can think of one – Triangulum – whose three main stars really do make a triangle. I suppose that with enough imagination you can picture a scorpion out of Scorpius or a crown out of Corona, but try making a dog out of Canis Minor or a goat out of Capricornus! Crux, the Southern Cross, is not X-shaped; it is more like a kite.

There used to be a smaller triangle, Triangulum Minor, near Aries, but it was one of the casualties of IAU revision.

The Centaur is a bright constellation in the southern sky. What was a centaur? Was it a dinosaur?

No. Dinosaurs existed. Centaurs never did – they are purely mythological. In the old legends a centaur was half-man, half-horse – a man down to the waist, and a horse below that.

Most centaurs were wild and unruly, but there was one notable exception: Chiron, the wise, kind teacher who cared for and taught many of the heroes of ancient Greek mythology. It is Chiron who is remembered by having one of the sky's brightest and most important constellations named in his honour.

If I go to Australia, will I be able to see the Pole Star?

No, you won't. As you know, the Earth spins on its axis once in about twenty-four hours. Northward, the axis points to the north pole of the sky – the North Celestial Pole, Pole Star, Polaris – Alpha Ursae Minoris, in the constellation of the Little Bear, lies very close to this point. So if you go to the North Pole, Polaris will be overhead, and the celestial equator will lie along the horizon. Where I live, at Selsey in Sussex, the latitude is roughly fifty-one degrees north, so that Polaris is fifty-one degrees above the horizon.

The further south you go, the lower Polaris sinks in the sky; where the latitude is zero degrees, the altitude of Polaris is zero degrees. Further south, it never rises at all. There is, of course, the South Celestial Pole, but unfortunately there is no bright star close to it. The South Pole Star is Sigma Octantis, in the constellation of the Octant, which is very dim – magnitude 5.25 – and not too easy to see with the naked eye; any mist or cloud will hide it. It does not even have an official proper name, though it is sometimes called Polaris Australis.

If I look at the stars on a clear night, I see a few which look red.
Why is this, and which are the stars?

Red stars have comparatively low surface temperatures – say around 3,000 degrees celsius, against well over 5,000 degrees for our yellow Sun. Red stars come in two classes; small, feeble dwarfs and huge, powerful giants. No red dwarfs are visible with the naked eye, so all those you have noticed have been giants.

Actually, only a few of these look obvious when observed without optical aid; colours do not show up – though with a telescope or binoculars they are striking. The most brilliant red stars are:

Betelgeux (Alpha Orionis): You cannot mistake this star, marking the Hunter's shoulder. Like many red supergiants it is variable, ranging officially from magnitude 0.3 to 0.7, but I have seen it at least equal to Rigel (0.1) and perhaps marginally brighter. When it is at minimum, it is comparable with Aldebaran. It is 13,000 times as luminous as the Sun.

Aldebaran (Alpha Tauri): Magnitude 0.9. This star is a giant, 140 times as luminous as the Sun, but not comparable with a supergiant. It is clearly orange-red, and is known as the 'Eye of the Bull'.

Antares (Alpha Scorpii): The 'Rival of Mars' in the Scorpion. This is fiery red. It is easy enough to find; there is a fainter star to either side of it. Like Betelgeux, Antares is a very powerful supergiant, 12,000 times as luminous as the Sun. Its magnitude is 1.1.

Other reddish or orange stars above the second magnitude are Gamma Crucis in the Southern Cross (too far south to be visible from Britain or Europe), Avior or Epsilon Carinae in the Keel (also too far south) and Alphard or Alpha Hydrae (2.0).

Do stars ever change colour?

Not over short periods – at least with regard to ordinary stars. For example, Arcturus will look orange tonight; it will still look orange tomorrow night, or in a year's time, or in AD 10,000. Eventually it will change, but not for tens of thousands of years. Stars like Arcturus, or the Sun, are pleasingly stable.

But there is one interesting story; it concerns Sirius, in Canis Major the Great Dog, and often known as the Dog Star. You can't overlook it whenever it is above the horizon, because it is so brilliant; it is more than half a magnitude brighter than its nearest rival, Canopus. If you are in any doubt, follow the line of Orion's Belt southwards (downwards, if you live in Britain) and there you will find Sirius, completely dominating the whole area. Of the planets, only Venus, Jupiter and (very occasionally) Mars can outshine it.

Sirius is a pure white star, which means that it has a hot surface; the temperature is over 10,000 degrees. When seen from Britain, when it is never really high above the horizon, its light comes to us after having passed through a deep layer of our unsteady atmosphere, and this makes it flash all the colours of the rainbow. All stars twinkle, but with Sirius this 'scintitlation' is particularly noticeable. I well remember looking at Sirius when I was in New Zealand; the Great Dog was overhead, and hardly twinkled at all. There was no trace of colour.

Now for the puzzle. A number of observers of Classical times listed stars which were obviously red, such as Betelgeux and Antares – and they included Sirius, even describing it as 'fiery'. They were quite definite about this: Sirius was red, and was often regarded as an unlucky star. Ptolemy, the greatest astronomer of ancient times, listed six bright stars which were red: Arctuus, Aldebaran, Pollux, Antares, Betelgeux, and Sirius. Of these, Pollux is 'off-white', Arcturus 'light orange', Aldebaran and Betetgeux 'orange-red', and Antares very red, so that the inclusion of Sirius is indeed curious. Ptolemy was writing about the year AD 140. Nothing could be more positive than that. So has Sirius really changed? Earlier there was a note by Seneca, who lived from 4 BC to AD 65: 'The redness of the Dog Star is more burning; that of Mars is milder, Jupiter is not red.' Yet Sirius has certainly not been red for the past thousand years, and in the tenth century AD the Arab astronomer Al-Sûfi wrote that it was pure white. So let us look at possible explanations.

Sirius is a Main Sequence star of spectral type A: it is shining in the same way as the Sun, by nuclear reactions. At the core, the nuclei of hydrogen atoms are combining to form nuclei of helium, releasing energy as they do so. Sirius shows no sign of using up its available hydrogen, and there seems little doubt that it has shone steadily for a very long time indeed. If we could accept a sudden change from red to white in historical times, it would mean that all our ideas about stellar evolution are wrong, and this seems to be out of the question.

One suggested explanation involves the companion of Sirius, known officially as Sirius B but often nicknamed the Pup. It was discovered in a rather unexpected way. Sirius is one of our nearest neighbours in the Galaxy; it is a mere 8.6 light-years away, and it shows definite 'proper motion' against the background of more distant stars – tiny, but quite measurable over a period of years. Around 1840 the German astronomer Friedrich Bessel found that Sirius was not moving in a straight line; it was 'weaving' its way along. Something was perturbing it, and Bessel realised that this must be due to a faint binary companion. He died unexpectedly in 1842 before he had organised a search, but in 1862 Alvan Clark, using the powerful telescope he had just completed, discovered the companion just where Bessel had expected it to be. It is not particularly faint (magnitude 8–6) but it is so drowned by the glare of Sirius A that it is distinctly elusive. It has only 1/10,000 the power of the primary. The pair are separated by 200 million miles on average (the orbit is somewhat eccentric), and the revolution period is fifty years.

Naturally the companion was expected to be cool, and red, but to everyone's surprise it turned out to be very hot; it looks dim simply because it is so small. It is about the same size as the Earth, but it is as massive as the Sun, so that it is incredibly heavy. Its density is well over 100,000 times that of water; the atoms in it are broken up and jam-packed, with almost no waste space. Stars of this kind are called white dwarfs. The Pup was the first to be found, but we now know that they are extremely common. They are well advanced in their evolution. A typical Main Sequence star uses up its store of hydrogen 'fuel', swells out to become a red giant and then collapses to become a white dwarf, eventually dying away altogether. (Of course this is a gross over-simplification, and a supergiant such as Betelgeux will explode as a supernova rather than subside quietly, but stars such as Sirius are not massive enough for this.) Yet since the Pup is now a white dwarf, it must have passed through the red giant stage, and there have been suggestions that this could solve the colour problem, so that in historic times it was the red-giant Pup which was the senior partner in the Sirius system. Unfortunately the time-scale is wrong; the change could not have been completed between Ptolemy's time and ours. Things do not happen as quickly as that.

Another idea is that the light reaching us from Sirius was temporarily reddened as it passed through a cloud of dust lying in the way. Again, possible – but if so, where is the cloud now?

No; my own view is that Sirius never was red, and the old observers were either misinterpreted or were just wrong. I find this difficult to believe in the case of Ptolemy, but all other explanations are even more difficult to believe, and the mystery must remain a mystery.

However, there are some stars which do show colour-changes: a nova is a formerly faint star which suddenly becomes brilliant for a few hours, days or weeks before fading back to obscurity. (Do not confuse ordinary novae with supernovae; the two classes are quite different.) One such nova was DQ Herculis, which flared up in 1934 and at its peak reached the first magnitude. I remember turning my telescope toward it when it was fading, and saw that it was vivid green. This did not last, and the nova has long since fallen below the reach of my telescope. But novae are exceptional; do not expect to go outdoors tomorrow night and find that Betelgeux has turned blue!

If I go outdoors tonight, will I be able to see the star known as the 'Wonderful Star'?

Only if you are lucky. The 'Wonderful Star' – Mira Ceti, to give it its proper name – is visible with the naked eye for only a few weeks in every year.

The story of its discovery is rather interesting. It begins in 1596 with the work of a Dutch amateur astronomer, Johann Fabricius, who noted an ordinary-looking star in the constellation of Cetus, the Whale. A few weeks later it had disappeared. It was seen again in 1603 by Johann Bayer, who was drawing up a new star catalogue in which he gave the stars in each constellation Greek letters, beginning with Alpha for the brightest star, Beta for the second, and so on, though it is true that this sequence is not always strictly followed. In Cetus, he catalogued a star of the third magnitude and gave it the letter Omicron; it was in exactly the same position as Fabricius' lost star. Again it vanished. Finally, in 1638, another Dutch astronomer, Phocytides Howarda, solved the mystery. Omicron Ceti is variable, and is at its brightest every 332 days. Without optical aid, it can be seen for only two or three weeks to either side of maximum. At minimum it drops to the tenth magnitude, so that even a small telescope will always show it. Its nickname of 'Mira Stelta', the Wonderful Star, is due to Hevelius, one of the leading telescopic observers of the early seventeenth century.

Both the period and the amplitude vary within narrow limits – the period by several days to either side of the mean of 332 days. At some maxima it may become no brighter than magnitude 4, but in 1779 it is said to become nearly as bright as Aldebaran. In February 1987 I estimated its peak magnitude as 2.3, comparable with the Pole Star. At least it is always easy to identify, because it has an M-type spectrum, and is obviously red.

There are now known to be many variable stars of the same type; we call them Mira stars. Quite a number are naked-eye objects when near maximum, for example Chi Cygni, R Leonis and R Hydrae – but Mira is the brightest of them. They are red giants, well advanced in their evolution, which is why they are unstable, and have started to pulsate. Mira itself is indeed a giant, with a diameter of around 600 million miles – large enough to contain not only the orbit of the Earth, but also that of Mars. It is 400 light-years from us, so that we now see it not as it is today, but as it used to be not long after the reign of Queen Elizabeth I.

It is not a solitary traveller in space. In 1923 R.G. Aitken, using the great thirty-six-inch refractor at the Lick Observatory in America, found a faint companion which proved to be small and hot. At around magnitude 9.5 it is not particularly faint, but is elusive because it is so close to the giant. The real separation is 6,500 million miles, and the orbital period is 400 years. An image taken with the Hubble Space Telescope shows that the giant is not spherical, but is shaped more like a rugger ball, undoubtedly because of the gravitational pull of the companion.

Go outdoors this evening and look for Mira by all means, if it is above the horizon. The map given here (see below) will show you where to find it. First look up your yearly handbook and check on the date of maximum; you may need binoculars or a telescope. You should not have much difficulty, and it is always pleasing to locate the 'Wonderful Star'.

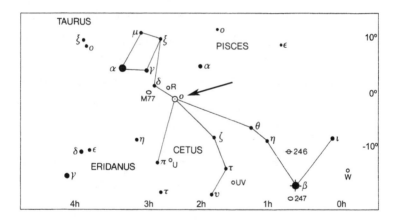

Mira Ceti

Have I a chance of seeing a supernova in our Galaxy?

It is always possible, because we never know when a supernova may appear. The last we saw in our Galaxy was Kepler's Star of 1604. By the law of averages, we are due for another. In 1987 there was a supernova in our satellite Galaxy (the Large Magellanic Cloud, which at its peak rose to the second magnitude.)

There are two promising supernova candidates. One is the erratic variable Eta Carinae, the far south of the sky (never visible from Britain or anywhere in mainland Europe). The other is Rho Cassiopeiae, in the far north, which never sets over Britain. It shines modestly close to Beta Cassiopeiae, one of the stars making up the enormous W pattern and since it is usually just above the fifth magnitude it is easy to locate; two stars of comparable brightness, Tau and Sigma Cassiopeiae, lie to either side of it. Rho does vary but seldom drops below naked-eye visibility (it last did so, briefly, in 2001).

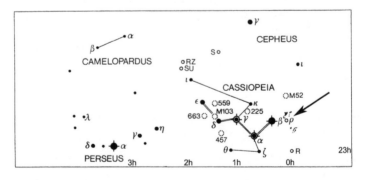

Rho (ρ) Cassiopeiae, between Sigma (σ) and Tau (τ), close to the second magnitude Beta (β)

Rho Cassiopeiae is a yellow hypergiant star – more powerful than an ordinary supergiant. It has 550,000 times the luminosity of the Sun, and its mass is probably about forty times that of the Sun; its huge globe could contain not only the orbit of the Earth, but also that of Mars; its distance is 11,600 light-years. It has used up its own store of nuclear fuel, and is drawing on its reserves; it has become unstable, and is bound to 'go supernova'. When that happens it will be a brilliant object in our sky, remaining prominent for months or even a year or two before the outburst is over.

Keep an eye on Rho Cassiopeiae. It looks innocent enough now, but at some stage it will briefly outshine all the other stars in the Milky Way. Do not be impatient; you may have to keep watching for a million years. We do not know when the outburst will happen, but happen it will.

I know that the nearest star is Alpha Centauri, just over four light-years away, but can there be any small stars which are closer than that?

Very unlikely. The sky is now so well mapped that we could hardly overlook a star which gave out any light. A star which had used up all its energy and was simply a cold, dead globe could exist only if it were too lightweight to make its presence felt by its gravitational pull. So I think that the Alpha Centauri system is really closer than any other star.

But what about a planet, which has broken away from its home system and has wandered off independently? This is quite possible, and we could not find it unless we were very lucky. There must be many comets, too, so that the space between the Solar System and Alpha Centauri will not be completely empty.

Why is the star Alpha Centauri so famous – and where can I see it in the sky?

It is the nearest star in the sky (not counting the Sun, of course), and in the night sky only Sirius and Canopus outshine it. Its magnitude is -0.27; it is one of only three stars above zero magnitude – the others are Sirius, Canopus, and (just) Arcturus. But if you live in England, as I do, you will never see it at all, because it is too far south. Its declination is approximately minus sixty degrees, which means that you will have to travel to an observing site on Earth south of forty degrees north. Alpha Centauri never rises over any part of Europe.

It has never had a universally accepted proper name, for reasons which I have never understood. One name is Toliman; another is Bundula; yet another is Rigil Kentaurus, which is not very suitable, as there is obvious danger of confusion with Rigel in Orion. Wartime air navigators (like me!) were expected to call it something, and on our RAF lists it was given as Rigel Kent. Astronomers in general simply refer to it as Alpha Centauri.

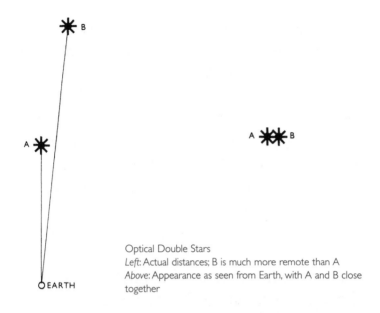

Optical Double Stars
Left: Actual distances; B is much more remote than A
Above: Appearance as seen from Earth, with A and B close together

It is not a single star; it is a wide, easy double, as any small telescope will show. The brighter member of the pair (A) is slightly larger and more massive than the Sun, and is 1.5 times as luminous; it is the same colour as the Sun. The secondary (B) is rather smaller and cooler than the Sun, and yellowish-orange. The two orbit their common centre of gravity in just over seventy-nine years, and the real distance between the two ranges between just over a thousand million miles out to 3.5 thousand million miles. Each is thought to be between five and six thousand million years old.

This is not all; there is a third, much fainter member of the system, known as Proxima because it is slightly closer to us, at a distance of 4.2 light-years. It is a dim red dwarf, about 13,000 astronomical units from the main pair. In our sky it is about two degrees from the main pair, and is by no means easy to identify unless you have a really good star map; it is only of the eleventh magnitude. It is probably orbiting the main pair, though there have been recent suggestions that it is merely 'passing by'.

Alpha Centauri lies near another bright star, Beta Centauri, and the two are often referred to as the Southern Pointers, because they show the way to the Southern Cross. But as so often happens, appearances are deceptive. Beta Centauri is an extremely luminous giant star, over 500 light-years away; it lies in the background, so to speak, and has absolutely no real connection with Alpha.

If I could manage to take a trip to Alpha Centauri, would I be able to see the Earth?

Yes, but you would need equipment much better than anything we can make today. Remember, you will be over four light-years away.

You could find the main constellations, because on the scale of the Galaxy the Sun and Alpha Centauri are not very far apart. The Sun would be a star of magnitude 0.5 (about as bright as Beta Centauri looks to us) near Epsilon Cassiopeiae, in the W pattern. Sirius would be in Orion, near Betelgeux.

But the Earth would be too faint, and too near the Sun, to be seen with any but a huge telescope; Jupiter would be easier. All in all, it does not seem likely that there are planets in the Alpha Centauri system – but one never knows. It is not impossible that at this very moment some astronomer there is looking across space at us.

Is it true that there is a star named Beetlejuice?

Well, some people call it that. As we've said, most of the star names (not all) are Arabic. The bright red star in Orion (the Hunter) marks his shoulder, and is called Betelgeux, but the Arabic alphabet is different from ours, and so the name is spelt in various ways – Betelgeuse and Betelgeuze, for instance. We do not know how it was originally pronounced – call it 'Beetlejuice' if you like; I prefer 'Bettle-gurz'. Opinionated scholars often claim to know the correct original pronunciations, but of course they don't. Take your pick.

If I went out to a planet in the system of Deneb, would I be able to see the Sun and the Earth?

Not without a powerful telescope. Deneb is a particularly powerful star; we are not sure of its exact distance, but it is probably around 3,000 light-years, and even though it is about 250,000 times as luminous as the Sun it is not even the brightest star in our sky – it only just makes the 'top twenty'. Whether it has a system of planets we are not certain, but I would say that the odds are rather against it, because as well as being so powerful it evolves much more quickly than our mild Sun and it will not go on shining for nearly as long. In a couple of million years' time the Sun will still be much the same as it is now, but Deneb will have exploded as a supernova.

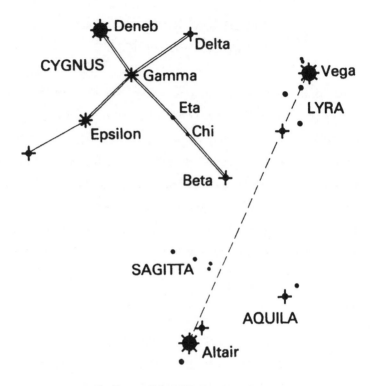

The 'Summer Triangle'; Deneb, Vega and Altair

But suppose it does have a planet, and that somehow or other we could pay it a visit? What would we see? In the daytime, Deneb would indeed be impressive, and probably very active. The night scene – that is to say, the view from the side of the planet turned away from Deneb – would be unfamiliar to us. Deneb is so far away, across the Galaxy, that we would see none of the constellations which grace our skies. There would be different stars, different patterns. But if we had the right equipment and the right telescope, we could locate the Sun as an excessively faint dot of light in a rich star-field. If sufficiently well equipped, we could even find out its spectrum, and learn that what we call the Sun is a totally undistinguished yellow dwarf. But it seems unlikely that even the planet Jupiter could be made out, and certainly not our tiny, close-in Earth. If they were skilful enough, astronomers there might well be able to measure the tiny 'wobbling' movement of the Sun caused by Jupiter's gravitational pull, but the pull of the Earth would surely be too slight to be measured even with instruments of amazing precision.

Are there planets? Are there any 'Denebian' astronomers? If so, what do they call the Sun? Will we ever be able to travel there? I doubt it, but I'd love to try!

Is there an owl in the sky?

Not one that hoots and flaps its wings – but there is the remarkable so-called Owl Nebula, No. 97 in Messier's famous catalogue of clusters and nebulae, and therefore known officially as M.97.

It is a planetary nebula, which, as we have noted, is not truly a nebula and has absolutely nothing to do with a planet. It is simply an old star which has used up its main store of nuclear fuel, and has expanded to become a giant; it then throws off its outer layers altogether, and these drift gradually away into space. For a brief period (brief, that is to say, cosmically speaking) the puffed-away shell of material can be seen round the old star, which is no longer a giant but has collapsed to become very small and very dense.

The most famous of these objects is the Ring Nebula in Lyra, which is a lovely sight in a telescope of moderate or large size, and has been likened to a small bicycle tyre with a central star. M.97, the Owl, is much fainter (its integrated magnitude is about 12) and its shape is different; there are two stars inside the ring, so that the appearance really does conjure up the picture of an owl's face. The apparent diameter is three minutes of arc, and the distance from us is about 10,000 light-years.

The Owl never sets over Britain or Europe, and lies in Ursa Major, the Great Bear, between Merak (the fainter of the two pointers to the Pole Star) and Phad, also a member of the pattern usually called the Plough (or, in America, the Dipper). The Owl's low surface brightness makes it rather elusive, but I have always found it easy enough with my 12.5-inch reflector. Seek it out by all means, but do not expect it to hoot!

Is there any danger than another star will move into the Solar System? If this did happen, what would be the effects on the Earth and the Sun?

The nearest star (Proxima Centauri) is over four light-years away, and if there were any luminous stars closer than that we would certainly have found them by now. Most of the stars are orbiting the centre of the Galaxy in very orderly paths, just as the Earth and the other planets orbit the Sun, so that the chances of an invasion of the Solar System at the present epoch are very low indeed. However, there are two points which have to be borne in mind. First, some stars orbit in eccentric and inclined paths, so that they cross the main streams – just as a car travelling in the right-hand lane of a busy road will have to cross the other lanes if he wants to take a slip-road on the opposite side. These stars are often called 'high-velocity' stars, but this is misleading; the stars are moving relatively quickly in comparison to the Earth, but not to the galactic centre. The chances of an encounter with one of these mavericks are remote, but in the long run we cannot say that they are absolutely nil.

Of course, there may be stars which we have not detected because they give out so little radiation, and we believe that a star such as the Sun will end up as a cold, dead globe, emitting no radiation at all. One of these dead or dying stars could approach the Sun, and once in the Solar System its gravitational pull would wreak havoc, perturbing the orbits of the planets so that the whole system would become unstable. Life on Earth would be very lucky to survive. The Sun would not be unharmed even if there were no direct collision, and a hit from even a relatively small, dead star would have catastrophic consequences. However, there is no reason to believe that an interstellar invader is anywhere near us – so to quote Corporal Jones, 'Don't panic!'

We have been talking about our present epoch, but things will not always be the same. Eventually our Galaxy will collide with the Andromeda Spiral, and the two galaxies will merge. All will be chaos; the stars of our Galaxy and the Andromeda will seldom meet head-on, but occasionally this must happen, and 'close encounters' will be much more common than they are today. But by then – over a thousand million years from now – changes in the Sun will have ended life on Earth, and the Earth itself may not have survived. At least mankind will have been given plenty of warning!

If I could go to a planet moving round a planet in the Andromeda Spiral, would I be able to see the Sun?

Yes, if you had the right equipment, but it would not be very easy, because the Sun is a very ordinary star – one of one hundred thousand million in the Milky Way system. The Andromeda Spiral is one and a half times larger than ours, and even so it is only just visible with the naked eye; it is over two million light-years away.

No doubt there are many planets there, and it is reasonable to believe that many of these must be of the same kind as Earth, fully able to support intelligent life. We have no proof of life elsewhere, but let us assume that it exists, and that we can travel to a planet orbiting a star in the Andromeda Spiral. What will the sky be like?

We will be able to make out our Galaxy, which is itself a spiral, though it will appear less bright than Andromeda does to us. Telescopes will be able to show individual stars, but our Sun, well away from the centre of our Galaxy, will be in no way distinguished, so that it will be rather like trying to identify one special sand grain in a full bucket – unless an Andromedan radio astronomer can pick up our radio signals, and locate us in that way. However, it takes a radio signal over two million years to travel from our Galaxy to Andromeda or vice versa, so that from Andromeda we are 'radio quiet' at the moment.

Have we ever transmitted a radio message to a star thousands of light-years away?

We have, though I must say that to me it seems rather pointless. Some years ago a signal was aimed at the globular cluster M.13 in Hercules, which is a huge symmetrical system containing hundreds of thousands of stars. Many of these stars are similar to the Sun, and no doubt have planetary systems, so that although we have no proof that life exists, there seems to be no reason why not. If so, and there are radio astronomers, they could pick up our message and realise that it must be artificial.

However, there is a problem. The Hercules cluster is 22,500 light-years away, and since radio waves travel at the same speed as light waves, our message will not arrive for 22,500 years. If some friendly Herculean operator picks it up and is courteous enough to reply promptly, we may hope to receive his reply in around the year AD 45200. Whether there will be anyone here to deal with it remains to be seen.

Will our Galaxy ever collide with another – and if so, will we be able to watch?

Yes. There is little doubt that our Galaxy will collide with the Andromeda Spiral, but we will not be able to watch, because it will not happen for at least a thousand million years, and by that time the Earth will no longer exist.

I had better begin by saying something about the Andromeda Spiral, M.31 (that is to say No. 31 in a famous catalogue of star-clusters and nebulae drawn up in 1781 by the French astronomer Charles Messier). It is a galaxy larger than the Milky Way system, and contains more than our quota of one hundred thousand million stars; it is spiral, like a cosmic Catherine-wheel. You can find it easily enough on a dark night, when it is above the horizon, because it is dimly visible with the naked eye, and is obvious in binoculars or any telescope. At first sight it is certainly not spectacular, and photographs taken with adequate telescopes are needed to show the spiral form really well. Unfortunately it lies at a narrow angle to us; if we could see it face-on (as with the Whirlpool Galaxy in the constellation of Canes Venatici, the Hunting Dogs) it would be truly magnificent.

For many years nobody knew how far away it was, and it was widely believed to be a minor feature of the Milky Way. This particular problem was solved by Edwin Hubble, in the early 1920s. Using the 100-inch reflector at Mount Wilson in California, then by far the most powerful telescope in the world, he studied some particular variable stars in the Spiral, known as Cepheids (because the best-known member of the class is Delta Cephei, in the far north of the sky). Cepheids brighten and fade regularly, in short periods; the period of Delta Cephei – that is to say the interval between successive maxima – is 5.3 days, while with another well-known Cepheid, Eta Aquilae in the Eagle, the period is just over 7 days. It had been found that the real luminosity of a Cepheid is linked with its period; the longer the period, the more powerful the star so that, for instance, Eta Aquilae is more powerful than Delta Cephei. Once you measure the period of a Cepheid, you know how luminous it is, and this means that you can find its distance. Of course there are many complications, such as the absorption of light in space, but in principle the method is straightforward enough – and it works.

When Hubble identified Cepheids in the Spiral, he realised that they were much too remote to be members of the Milky Way system. It followed that the Andromeda Spiral was a galaxy in its own right, and the same would be true of the other spirals and what had been called 'starry nebulae'. The universe was much larger than most people had believed – yet the Andromeda Galaxy is the nearest of all the spirals, and is the only one clearly visible without optical aid. Hubble gave the distance as 750,000 light-years. This has proved to be an under-estimate, and the modern value is about 2.3 million light-years. Look at the Spiral tonight, and you will see it not as it is now, but as it used to be in the days long before dinosaurs roamed the world.

The Spiral is at a safe distance, but it is rushing toward us, and eventually the two systems are bound to collide. By then – over a thousand million years in the future – our Sun will have become much more luminous than it is now, and even if the Earth survives, which does not seem likely, it will have become a molten, uninhabitable mass. Our Galaxy and M.31 will not immediately meet head-on, and there will be a long period when they will, so to speak, waltz round each other, but eventually the two systems will begin to merge. The individual stars are widely separated, and this means that direct collisions will be very rare, but the gas and dust between the stars will be colliding all the time, and the effects will be profound. Pressure waves will trigger off supernova explosions and frenetic outbursts of star formation; chaos will ensue, and some stars (perhaps even what may be left of the Sun!) may be driven away. The collision will not be a brief affair, and will go on for perhaps a thousand million years, but eventually the graceful spiral forms will be destroyed, and the end product is likely to be a single large elliptical galaxy.

This may sound speculative, but it is based on firm evidence, because we can see what is happening elsewhere in the universe. There are vast numbers of ellipticals which are presumably due to spiral mergers, and relatively near at hand we can see the Antennae Galaxies in the constellation of Corvus (the Crow) which are merging at the present time. We can see how both have been distorted, and the long streamers of gas and dust, studded with stars, do conjure up a picture of ants' antennae!

We can learn a good deal from the Antennae which are forty-six million light-years away (closer than was believed until very recently). Just over a thousand million years ago they were separate, one of them a normal spiral and the other a barred spiral, where the arms extend from a bar right through the system. About 900 million years ago they approached each other, and 600 million years ago they passed through each other. In about 400 million years time the nuclei of the two old galaxies will meet and merge, leaving a single elliptical. Our Galaxy and M.31 are larger and more massive than the Antennae, but the sequence of events will follow the same pattern, though it will take longer. Civilisations living on planets in our Galaxy must then be prepared for exciting developments in their skies!

If I could be around when the Andromeda Galaxy collides with the Milky Way, what would I see?

Nothing immediately spectacular, but it would certainly be an interesting period!

All groups of galaxies are receding from all other groups, but the Andromeda Galaxy and our Galaxy are both members of the Local Group, and at present they are rushing toward each other. Eventual collision is inevitable, though not for at least a thousand million years, and by then all life on Earth will have become extinct; the Earth itself may not survive. As the sun ages and uses up its hydrogen 'fuel', it is getting hotter – much too slowly for us to notice, but inexorably. A thousand million years hence the Earth will have become too hot for life to survive, and the oceans will boil away. When the Sun becomes a red giant, the Earth will either be destroyed or else turned into a seething, molten mass.

But what about the collision? It will not happen suddenly, and first the two Galaxies will waltz round each other. Then, at last, they will meet, but individual stars will seldom hit each other, and the effect will be much the same as two orderly crowds moving through each other in different directions. However, the material between the stars will be colliding all the time, and there will be tremendous disturbances, with bursts of star formation. This will continue throughout the collision, which will last for a very long time indeed – certainly at least a thousand million years. At last the cores of the two galaxies, with their black holes, will merge. The graceful spiral forms will be destroyed, and the end product is likely to be a single giant elliptical system.

Whether there will be any life by then, either in our Galaxy or in M.31, we have no means of knowing; we can only speculate, and it is always possible that our whole concept is wrong. However, we can see galaxy collisions going on – with the Antennae, for example, and the Mice – and we are fairly sure that this really will be the sequence of events.

What is dark matter, and who discovered it?

If I could answer that question, I would be Astronomer Royal! But I can't, and at the moment neither can anybody else. We are ninety-nine per cent certain that dark matter exists, but we have absolutely no idea about its nature, and we can detect it only by its effects upon visible matter. The original work on it was due to Fritz Zwicky, one of the most extraordinary astronomers of modern times. He was born in 1898, and died in 1974. He was Swiss, but lived for most of his life in the USA, and was able to make use of the Mount Wilson and Palomar reflectors, then the most powerful telescopes in the world.

Zwicky paid great attention to stellar evolution, and realised that when a very massive star runs out of nuclear 'fuel' it will explode in a colossal outburst; these are what Zwicky called supernovae, a term which of course is still in use today. Supernovae are not common, and over the past thousand years only three have been seen in our Galaxy: in 1006, 1572 and 1604. All these became so brilliant that they were visible with the naked eye in broad daylight. They faded away after several months, but we can identify their remnants: the Crab Nebula in Taurus, with its pulsar, is the wreck of the supernova of 1006.

These outbursts are immensely powerful; at its peak a supernova may equal the combined luminosity of all the other stars in its host galaxy. This means that supernovae can be seen thousands of millions of light-years away, and they are observed frequently; amateurs are invaluable here – for example the English amateur Tom Boles has so far made more than a hundred discoveries. Because supernovae of a certain type peak at the same luminosity, they can be used as 'standard candles' out into the far reaches of the universe. Zwicky calculated that in every galaxy there should be at least one supernova every 200 years or so. Few people believed him, but with great difficulty he managed to get permission to make a search with the 100-inch Hooker reflector at Mount Wilson, and it did not take him long to prove his point. By

1936 he had discovered thirty supernovae in external galaxies. It was also Zwicky who first claimed that a supernova could end up as a tiny, incredibly dense globe made up of neutrons. Again he was right; the Crab pulsar is a rapidly-rotating neutron star. A supernova of even greater mass may end its career by producing a Black Hole.

Zwicky next turned to the ways in which stars and galaxies move. It was known that galaxies form clusters, and our own Galaxy is a member of such a cluster: the Local Group, made up of three large galaxies (the Andromeda Spiral M.31, our Galaxy and the Triangulum Spiral M.33), several systems of moderate size (such as the Magellanic Clouds) and at least thirty dwarfs. Other clusters contain hundreds or even thousands of galaxies, all of which are moving around. Each cluster is racing away from each other cluster, so that the whole universe is expanding. But Zwicky also realised something else. Inside any particular cluster, the galaxies are moving so quickly relative to each other that they ought to fly apart, and the cluster would disperse and lose its identity – but this does not happen. Something is 'glueing' them together, and this must exert enough gravitational pull to prevent the cluster from dispersing. Zwicky called it missing mass; we call it dark matter.

But what exactly is it? Can it be due to large numbers of stars so dim that we cannot see them? Is it matter locked up inside Black Holes? Do neutrinos, strange particles, exist in vast numbers? Or are we dealing with material so strange and so alien that we cannot detect it with any scientific equipment yet built or even planned? This is the favoured answer at present, but to me at least it seems to be simply a fudge. We know no more about the nature of dark matter now than Zwicky did.

There is also another point. Galaxies such as ours are rotating; it takes the Sun 225 million years to complete one orbit round the galactic centre (a period often referred to as the cosmic year). A star closer to the centre would be expected to move more quickly – in fact to have a greater orbital velocity. This is what happens in the Solar System; Mercury, the closest-in planet, has the greatest orbital velocity while Neptune, the furthest-out, is the slowest. In rotating galaxies, the situation is different, and the orbital velocities of stars do not decrease with increasing distance from the centre. There is only one possible answer. Over ninety-nine per cent of the total mass of the Solar System is concentrated in the central body – the Sun – but the mass of a galaxy is not concentrated in the centre. It is spread throughout the whole system, even though we cannot see it. Again we have to deal with invisible dark matter.

All this shows that Fritz Zwicky was a brilliant research astronomer, and undeniably he was often right when almost everybody else was wrong. But to say that he was eccentric is to put it mildly. I met him once, when he was Professor of Astrophysics at the California Institute of Technology, but all in all I doubt whether he and I would have had much in common. He had a very 'short fuse', and anyone who disagreed with him was automatically classed as a mortal enemy. Luckily my one encounter with him was uneventful enough; after all, he was one of the world's greatest astrophysicists, whereas I was a mere amateur Moon-mapper!

He was physically very strong, and to demonstrate this he made a habit of doing handstands in the dining hall of the observatory during dinner. He was convinced that others were stealing his ideas without giving him due credit – on one occasion he ordered his night assistant to open the slit of the observatory dome and fire bullets through it, which might improve the seeing conditions (it didn't). But it was with his colleague Walter Baade that he really excelled himself. The two were collaborating on some important researches; Baade was German, though he had spent many years in America. Zwicky referred to him as a Nazi, and threatened to kill him if he ever found him alone on the observatory campus. Baade took this seriously, particularly as Zwicky's appearance was often described as 'menacing'. When Zwicky retired in 1969, it is not likely that many of his colleagues were sorry to see him go, but there can be no doubt that his work was of the greatest value to those who succeeded him.

Is it true that the cosmic microwave background radiation was once thought to be due to pigeons?

Quite true. We believe that the universe was created with the Big Bang, 13.7 thousand million years ago. It was then incredibly hot, and radiation spread out in all directions. This expansion diluted the radiation, and the wavelength of the light increased. It was calculated that the background radiation should still be detectable and in 1965 the American scientist R. Dicke suggested that by now its temperature should have fallen to about three degrees above absolute zero – that is to say – 270 degrees celcius.

Meanwhile, A. Penzias and R. Wilson, two leading US radio astronomers, had built a special radio antenna for various investigations they were carrying out; the antenna was horn-shaped. The investigations went ahead, but all the tine there was an irritating background 'hiss' that they could not identify. Try as they might, they could not associate it with anything at all. Then they had an idea. Nearby pigeons had made a habit of leaving their droppings in the horn antenna; could these be the cause of the trouble?

With considerable difficulty the droppings were cleaned away, and the pigeons persuaded to relieve themselves elsewhere. No use! The hiss was still there and the two radio astronomers had to admit that they were baffled.

Robert Dicke heard about the problem – and straightaway found the answer. The pigeons were innocent; what the horn antenna was picking up was nothing more nor less than the three degree cosmic background radiation – the last 'echo' of the Big Bang.

This proved to be one of the most important astronomical discoveries since Hubble had proved that the 'spiral nebulae' really are independent star-systems, not minor members of our Galaxy. No doubt the pigeons were subsequently left unmolested.

At the moment of the Big Bang when the universe was created, was there really a loud bang?

Not in the usual meaning of 'bang'. The term was first used contemptuously by Sir Fred Hoyle, who never believed in it, but it is decidedly misleading.

According to most astronomers, the universe came into existence suddenly, 13.7 thousand million years ago, but not with an explosion in space. Before the creation, there was no space – there was nothing at all, because space, time and matter came into existence at the same moment. We cannot ask what happened before that, because there was no 'before'.

Does this confuse you? I expect so – it certainly confuses me, and I am quite sure that it confuses everybody else, even our greatest scientists. I think we have to admit that our brains are unable to cope with problems of this sort; we simply cannot understand them. And we are equally at a loss if we try to explain how and why the so-called Big Bang happened.

Of course, there are some scientists – not many, now – who believe that the whole concept is wrong, that the universe has always existed, and that it will exist forever. This certainly does away with the Big Bang and no 'before', but can you imagine a period of time that has no beginning? No – and we are no better off.

Incidentally, if there really had been a thunderous crash at the instant of the Big Bang, it would have passed unnoticed. There would have been nobody around to hear it!

Will the universe ever come to an end?

We do not know. According to the Big Bang theory, now accepted by most astronomers (not all), everything came suddenly into existence 13.7 thousand million years ago, and this is when 'time' started; as I have already said (p. 216), there was no 'before'. Therefore, if the universe comes to an end, 'time' will end too, and there will be no 'after'.

The sequence of events sounds rather depressing. Over the ages, stars die until all the galaxies are dark. Eventually atoms themselves disintegrate, and we are left with nothing but radiation; presumably even this would be lost if 'time' came to an end. Nothing at all would remain. This kind of concept is too much for our brains.

There is another theory which leads to a very different ending. We know that all the groups of galaxies are racing away from each other, so that the universe is expanding. If the rate of expansion were to increase sufficiently, the universe would, to all intents and purposes, explode. Or will gravity eventually overcome the force of expansion, so that the galaxies will rush together and meet in a Big Crunch around eighty thousand million years hence? And if so, will this be succeeded by another Big Bang, so that the entire cycle will be repeated? This is what I call the Concertina Universe: Bang, Crunch, Bang, Crunch – could this perhaps be repeated for ever?

I do not know. Nobody knows. But one thing I do know, without a shadow of doubt: if the universe does come to an end, you and I won't be there to see it!

INDEX